T0201327

Data Science: The Executive Summary

Data Science: The Executive Summary

A Technical Book for Non-Technical Professionals

Field Cady

This edition first published 2021
© 2021 by John Wiley & Sons, Inc.

Registered Office
John Wiley & Sons, Inc., 111 River Street, Hoboken, NJ 07030, USA

Editorial Office
111 River Street, Hoboken, NJ 07030, USA

For details of our global editorial offices, customer services, and more information about Wiley products visit us at www.wiley.com.

Wiley also publishes its books in a variety of electronic formats and by print-on-demand. Some content that appears in standard print versions of this book may not be available in other formats.

Library of Congress Cataloging-in-Publication Data

Names: Cady, Field, 1984- author.
Title: Data science : the executive summary : a technical book for
   non-technical professionals / by Field Cady.
Description: Hoboken, NJ : Wiley, 2021. | Includes bibliographical
   references and index.
Identifiers: LCCN 2020024708 (print) | LCCN 2020024709 (ebook) | ISBN
   9781119544081 (hardback) | ISBN 9781119544166 (adobe pdf) | ISBN
   9781119544173 (epub)
Subjects: LCSH: Data mining.
Classification: LCC QA76.9.D343 C33 2021 (print) | LCC QA76.9.D343
   (ebook) | DDC 006.3/12–dc23
LC record available at https://lccn.loc.gov/2020024708
LC ebook record available at https://lccn.loc.gov/2020024709

Cover Design: Wiley
Cover Image: © monsitj/Getty Images

Set in 9.5/12.5pt STIXTwoText by SPi Global, Chennai, India

Printed in the United States of America

SKY10022856_112520

For my Uncle Steve, who left the world on the day this book was finished. And for my son Cyrus, who entered shortly thereafter.

Contents

1

Introduction

1.1 Why Managers Need to Know About Data Science

There are many "data science for managers" books on the market today. They are filled with business success stories, pretty visualizations, and pointers about what some of the hot trends are. That material will get you rightfully excited about data science's potential, and maybe even get you started off on the right foot with some promising problems, but it isn't enough to see projects over the finish line or bring the full benefits of data to your organization. Depending on your role you may also need to decide how much to trust a piece of analytical work, make final calls about what tools your company will invest in, and hire/manage a team of data scientists. These tasks don't require writing your own code or performing mathematical derivations, but they do require a solid grounding in data science concepts and the ability to think critically about them.

In the past, mathematical disciplines like statistics and accounting solved precisely defined problems with a clear business meaning. You don't need a STEM degree to understand the idea of testing whether a drug works or balancing a checkbook! But as businesses tackle more open-ended questions, and do so with datasets that are increasingly complex, the analytics problems become more ambiguous. A data science problem almost never lines up perfectly with something in a textbook; there is always a business consideration or data issue that requires some improvisation. Flexibility like this can become recklessness without fluency in the underlying technical concepts. Combine this with the fact that data science is fast becoming ubiquitous in the business world, and managers and executives face a higher technical bar than they ever did in the past.

Business education has not caught up to this new reality. Most careers follow a "business track" that teaches few technical concepts, or a "technical track" that focuses on hands-on skills that are useless for businesspeople. This book charts a middle path, teaching non-technical professionals the core concepts of modern data science. I won't teach you the brass tacks of how to do the work yourself (that

Data Science: The Executive Summary - A Technical Book for Non-Technical Professionals,
First Edition. Field Cady.
© 2021 John Wiley & Sons, Inc. Published 2021 by John Wiley & Sons, Inc.

truly is for specialists), but I will give you the conceptual background you need to recognize good analytics, frame business needs as solvable problems, manage data science projects, and understand the ways data science is likely to transform your industry.

In my work as a consultant I have seen PMs struggle to mediate technical disagreements, ultimately making decisions based on peoples' personalities rather than the merits of their ideas. I've seen large-scale proof-of-concept projects that proved the wrong concept, because organizers set out inappropriate success metrics. And I've seen executives scratching their heads after contractors deliver a result, unable to see for themselves whether they got what they paid for.

Conversely, I have seen managers who can talk shop with their analysts, asking solid questions that move the needle on the business. I've seen executives who understand what is and isn't feasible, instinctively moving resources toward projects that are likely to succeed. And I've seen non-technical employees who can identify key oversights on the part of analysts and communicate results throughout an organization.

Most books on data science come in one of two types. Some are written for aspiring data scientists, with a focus on example code and the gory details of how to tune different models. Others assume that their readers are unable or unwilling to think critically, and dumb the technical material down to the point of uselessness. This book rejects both those approaches. I am convinced that it is not just possible for people throughout the modern business workforce to learn the language of data: it is essential.

1.2   The New Age of Data Literacy

Analytics used to play a minor role in business. For the most part it was used to solve a few well-known problems that were industry-specific. When more general analytics *was* needed, it was for well-defined problems, like conducting an experiment to see what product customers preferred.

Two trends have changed that situation. The first is the intrusion of computers into every aspect of life and business. Every phone app, every new feature in a computer program, every device that monitors a factory is a place where computers are making decisions based on algorithmic rules, rather than human judgment. Determining those rules, measuring their effectiveness, and monitoring them over time are inherently analytical. The second trend is the profusion of data and machines that can process it. In the past data was rare, gathered with a specific purpose in mind, and carefully structured so as to support the intended analysis. These days every device is generating a constant stream of data, which is passively gathered and stored in whatever format is most convenient. Eventually it gets used by high-powered computer clusters to answer a staggering range of questions, many of which it wasn't really designed for.

I don't mean to make it sound like computers are able to take care of everything themselves – quite the opposite. They have no real-world insights, no creativity, and no common sense. It is the job of humans to make sure that computers' brute computational muscle is channeled toward the right questions, and to know their limitations when interpreting the answers. Humans are not being replaced – they are taking on the job of shepherding machines.

I am constantly concerned when I see smart, ethical business people failing to keep up with these changes. Good managers are at risk of botching major decisions for dumb reasons, or even falling prey to unscrupulous snake oil vendors. Some of these people are my friends and colleagues. It's not a question of intelligence or earnestness – many simply don't have the required conceptual background, which is understandable. I wrote this book for my friends and people like them, so that they can be empowered by the age of data rather than left behind.

1.3   Data-Driven Development

So where is all of this leading? Cutting out hyperbole and speculation, what does it look like for an organization to make full use of modern data technologies and what are the benefits? The goal that we are pushing toward is what I call "data-driven development" (DDD). In an organization that uses DDD, all stages in a business process have their data gathered, modeled, and deployed to enable better decision making. Overall business goals and workflows are crafted by human experts, but after that every part of the system can be monitored and optimized, hypotheses can be tested rigorously and retroactively, and large-scale trends can be identified and capitalized on. Data greases the wheels of all parts of the operation and provides a constant pulse on what's happening on the ground.

I break the benefits of DDD into three major categories:

1. *Human decisions are better-informed*: Business is filled with decisions about what to prioritize, how to allocate resources, and which direction to take a project. Often the people making these calls have no true confidence in one direction or the other, and the numbers that could help them out are either unavailable or dubious. In DDD the data they need will be available at a moment's notice. More than that though, there will be an understanding of how to access it, pre-existing models that give interpretations and predictions, and a tribal understanding of how reliable these analyses are.
2. *Some decisions are made autonomously*: If there is a single class of "killer apps" for data science, it is machine learning algorithms that can make decisions

without human intervention. In a DDD system large portions of a workflow can be automated, with assurances about performance based on historical data.

3. *Everything can be measured and monitored*: Understanding a large, complex, real-time operation requires the ability to monitor all aspects of it over time. This ranges from concrete stats – like visitors to a website or yield at a stage of a manufacturing pipeline – to fuzzier concepts like user satisfaction. This makes it possible to constantly optimize a system, diagnose problems quickly, and react more quickly to a changing environment.

It might seem at first blush like these benefit categories apply to unrelated aspects of a business. But in fact they have much in common: they rely on the same datasets and data processing systems, they leverage the same models to make predictions, and they inform each other. If an autonomous decision algorithm suddenly starts performing poorly, it will prompt an investigation and possibly lead to high-level business choices. Monitoring systems use autonomous decision algorithms to prioritize incidents for human investigation. And any major business decision will be accompanied by a plan to keep track of how well it turns out, so that adjustments can be made as needed.

Data science today is treated as a collection of stand-alone projects, each with its own models, team, and datasets. But in DDD all of these projects are really just applications of a single unified system. DDD goes so far beyond just giving people access to a common database; it keeps a pulse on all parts of a business operation, it automates large parts of it, and where automation isn't possible it puts all the best analyses at people's fingertips.

It's a waste of effort to sit around and guess things that can be measured, or to cross our fingers about hypotheses that we can go out and test. Ideally we should spend our time coming up with creative new ideas, understanding customer needs, deep troubleshooting, or anticipating "black swan" events that have no historical precedent. DDD pushes as much work as possible onto machines and pre-existing models, so that humans can focus on the work that only a human can do.

1.4 How to Use this Book

This book was written to bring people who don't necessarily have a technical background up to speed on data science. The goals are twofold: first I want to give a working knowledge of the current state of data science, the tools being used, and where it's going in the foreseeable future. Secondly, I want to give a solid grounding in the core concepts of analytics that will never go out of date. This book may also be of interest to data scientists who have nitty-gritty technical chops but want to take their career to the next level by focusing on work that moves the business needle.

The first part of this book, The Business Side of Data Science, stands on its own. It explains in non-technical terms what data science is, how to manage, hire, and work with data scientists, and how you can leverage DDD without getting into the technical weeds.

To really achieve data literacy though requires a certain amount of technical background, at least at a conceptual level. That's where the rest of the book comes in. It gives you the foundation required to formulate clear analytics questions, know what is and isn't possible, understand the tradeoffs between different approaches, and think critically about the usefulness of analytics results. Key jargon is explained in basic terms, the real-world impact of technical details is shown, unnecessary formalism is avoided, and there is no code. Theory is kept to a minimum, but when it is necessary I illustrate it by example and explain why it is important. I have tried to adhere to Einstein's maxim: "everything should be made as simple as possible... but not simpler."

Some sections of the book are flagged as "advanced material" in the title. These sections are (by comparison) highly technical in their content. They are necessary for understanding the strengths and weaknesses of specific data science techniques, but are less important for framing analytics problems and managing data science teams.

I have tried to make the chapters as independent as possible, so that the book can be consumed in bite-sized chunks. In some places the concepts necessarily build off of each other; I have tried to call this out explicitly when it occurs, and to summarize the key background ideas so that the book can be used as a reference.

2

The Business Side of Data Science

A lot of this book focuses on teaching the analytics concepts required to best lever-age data science in your organization. This first part, however, zooms out and takes the "pure business" perspective. It functions as a primer on what value data scien-tists can bring to an organization, where they fit into the business ecosystem, and how to hire and manage them effectively.

2.1 What Is Data Science?

There is no accepted definition of "data science." I don't expect there to ever be one either, because its existence as a job role has less to do with clearly defined tasks and more to do with historical circumstance. The first data scientists were solv-ing problems that would normally fall under the umbrella of statistics or business intelligence, but they were doing it in a computationally-intensive way that relied heavily on software engineering and computer science skills. I'll talk more about these historical circumstances and the blurry lines between the job titles shortly, but for now a good working definition is:

Data Science: *Analytics work that, for one reason or another, requires a substan-tial amount of software engineering skills*

This definition technically applies to some people who identify as statisticians, business analysts, and mathematicians, but most people in those fields can't do the work of a good data scientist. That might change in the future as the educational system catches up to the demands of the information economy, but for the time being data scientists are functionally speaking a very distinctive role.

2.1.1 What Data Scientists Do

Data science can largely be divided into two types of work: the kind where the clients are humans and the kind where the clients are machines. These styles are

Data Science: The Executive Summary - A Technical Book for Non-Technical Professionals,
First Edition. Field Cady.
© 2021 John Wiley & Sons, Inc. Published 2021 by John Wiley & Sons, Inc.

often used on the same data, and leverage many of the same techniques, but the goals and final deliverables can be wildly different.

If the client is a human, then typically you are either investigating a business situation (what are our users like?) or you are using data to help make a business decision (is this feature of our product useful enough to justify the cost of its upkeep?). Some of these questions are extremely concrete, like how often a particular pattern shows up in the available data. More often though they are open-ended, and there is a lot of flexibility in figuring out what best addresses the business question. A few good examples would be

- Quantifying the usefulness of a feature on a software product. This involves figuring out what "success" looks like in the data logs, whether some customers are more important than others, and being aware of how the ambiguities here qualify the final assessment.
- Determining whether some kind of compelling pattern exists in the available data. Companies are often sitting on years' worth of data and wondering whether there are natural classes of users, or whether there are leading indicators of some significant event. These kinds of "see what there is to see" analyses are often pilots, which gauge whether a given avenue is worth pouring additional time and effort into.
- Finding patterns that predict whether a machine will fail or a transaction will go through to completion. Patterns that correlate strongly with failure may represent problems in a company's processes that can be rectified (although they could equally well be outside of your control).
- Test which of two versions of a website works better. AB testing like this largely falls under the domain of statistics, but measuring effectiveness often requires more coding than a typical statistician is up for.

Typically the deliverables for this kind of work are slide decks, written reports, or emails that summarize findings.

If the client is a machine then typically the data scientist is devising some logic that will be used by a computer to make real-time judgements autonomously. Examples include

- Determining which ad to show a user, or which product to try up-selling them with on a website
- Monitoring an industrial machine to identify leading indicators of failure and sound an alarm when a situation arises
- Identifying components on an assembly line that are likely to cause failures downstream so that they can be discarded or re-processed

In these situations the data scientist usually writes the final code that will run in a production situation, or at least they write the first version of it that is

Table 2.1 Data science work can largely be divided into producing human-understandable insights or producing code and models that get run in production.

Client	Uses	Deliverables	Special considerations
Human	• Understanding the business • Helping humans make data-driven decisions	• Slide decks • Narratives that explain how/why	• Formulating questions that are useful and answerable • How to measure business outcomes
Machine	• Making decisions autonomously	• Production code • Performance specs	• Code quality • Performance in terms of speed

incorporated by engineers into the final product. In some cases they do not write the real-time code, but they write code that is periodically re-run to tune the parameters in the business logic.

The differences between the two categories of data science are summarized in Table 2.1. This might seem like I'm describing two different job roles, but in fact a given data scientist is likely to be involved in both types of work. They use a lot of the same datasets and domain expertise, and there is often a lot of feedback between them. For example, I mentioned identifying leading indicators of failure in an industrial machine – these insights can be used by a human being to improve a process or by a machine to raise an alarm when failure is likely imminent.

Another common case, which does not fit cleanly into the "client is a machine" or "client is a human" category, is setting up analytics infrastructures. Especially in small organizations or teams, a data scientist often functions as a one-person shop. They set up the databases, write the pre-processing scripts that get the data into a form where it is suitable for a database, and create the dashboards that do the standard monitoring. This infrastructure can ultimately be used to help either human or machine clients.

2.1.2   History of Data Science

There are two stories that come together in the history of data science. The first concerns the evolution of data analysis methodology and statistics, and especially how it was impacted by the availability of computers. The second is about the data itself, and the Big Data technologies that changed the way we see it.

Data has been gathered and analyzed in some form for millennia, but statistics as a discipline is widely dated to 1662. In that year John Graunt and William Petty used mathematical methods to study topics in demographics, like life expectancy tables and the populations of cities. Increasingly mathematical techniques were developed as the scientific revolution picked up steam, especially in the context of using astronomical data to estimate the locations and trajectories of celestial bodies. People began to apply these scientific techniques to other areas, like Florence Nightingale in medicine. The Royal Statistical Society was founded in 1834, recognizing the general-purpose utility of statistics. The greatest figure in classical statistics was Ronald Fisher. In the early twentieth century, he almost single-handedly created the discipline in its modern form, with a particular focus on biological problems. The name of the game was the following: distill a real-world situation down into a set of equations that capture the essence of what's going on, but are also simple enough to solve by hand.

The situation changed with the advent of computers, because it was no longer necessary to do the math by hand. This made it possible to try out different analytical approaches to see what worked, or even to just explore the data in an open-ended way. It also opened the door to a new paradigm – the early stages of machine learning (ML) – which began to gain traction both in and especially outside of the statistics community. In many problems the goal is simply to make accurate predictions, by hook or crook. Previously you did that by understanding a real-world situation and distilling it down to a mathematical model that (hopefully!) captured the essence of the phenomena you were studying. But maybe you could use a very complicated model instead and just solve it with a computer. In that case you might not need to "understand" the world: if you had enough data, you could just fit a model to it by rote methods. The world is a complex place, and fitting complicated models to large datasets might be more accurate than any idealization simple enough to fit into a human brain.

Over the latter part of the twentieth century, this led to something of a polarization. Traditional statisticians continued to use models based on idealizations of the world, but a growing number of people experimented with the new approach. The divide was best described by Leo Breiman – a statistician who was one of the leading lights in ML – in 1991:

> There are two cultures in the use of statistical modeling to reach conclusions from data. One assumes that the data are generated by a given stochastic data model. The other uses algorithmic models and treats the data mechanism as unknown. The statistical community has been committed to the almost exclusive use of data models. This commitment has led to irrelevant theory, questionable conclusions, and has kept statisticians from working on a large range of interesting current problems. Algorithmic modeling, both in theory and practice, has developed rapidly in fields outside statistics. It can be used both on large complex data sets and as a more accurate and informative alternative to data modeling on smaller data sets. If our goal as a field is to use data to solve problems, then

we need to move away from exclusive dependence on data models and adopt a more diverse set of tools.

The early "algorithmic models" included the standard classifiers and regressors that the discipline of ML is built on. It has since grown to include deep learning, which is dramatically more complicated than the early models but also (potentially) much more powerful.

The story of the data itself is much more recent. Three years after Breiman's quote, in 2004, the Big Data movement started when Google published a paper on MapReduce, a new software framework that made it easy to program a cluster of computers to collaborate on a single analysis, with the (potentially very large) dataset spread out across the various computers in the cluster. Especially in the early days clusters were notoriously finicky, and they required far more IT skills than a typical statistician was likely to know.

The other thing that was new about Big Data was that the data was often "unstructured." This means it wasn't relevant numbers arranged into nice rows and columns. It was webpages in HTML, Word documents, and log files belched out by computers – a cacophony of different formats that were not designed with any particular analytics question in mind. Companies had been sitting of backlogs of messy legacy data for years, and MapReduce finally made it possible to milk insights out of them.

Most of the labor of Big Data was in just getting the data from its raw form into tables of numbers where statistics or ML could even be applied. This meant large code frameworks for handling all the different formats, and business-specific knowledge for how to distill the raw data into meaningful numbers. The actual statistics was often just fitting a line – you need some mathematical knowledge to do this responsibly given so many moving parts, but it hardly requires a full-blown statistician.

And so the hybrid role of data scientist was born. They were mostly drawn from the ranks of computer scientists and software engineers, especially the ones who were originally from math-heavy backgrounds (I've always been shocked by how many of the great computer scientist were originally physicists). This gave rise to the mythology of the data scientist – a brilliant polymath who had a mastery of all STEM knowledge. After all, look at the resumes of the people who became data scientists!

The reality though was that you just needed somebody who was competent at both coding and analytics, and the people who knew both of these largely unrelated disciplines were predominantly polymaths. It doesn't need to be this way, and the educational system is quickly figuring out that solid coding skills are important for people in all STEM fields. In explaining my work, I often jokingly tell people

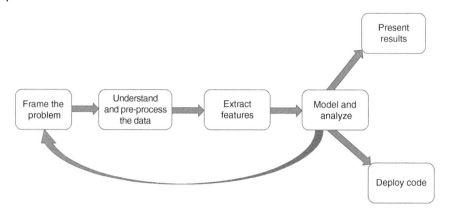

Figure 2.1 The process of data science is deeply iterative, with the questions and methods evolving as discoveries are made. The first stage – and the most important – is to ask questions that provide the most business value.

that "if you take a mediocre computer programmer and a mediocre statistician and put them together you get a good data scientist."

Technology evolved. As individual computers became more powerful there was less need to resort to cluster computing (which has many disadvantages I will discuss later in the book – you should only use a cluster if you really have to). Cluster computing itself improved, so that data scientists could spend less time troubleshooting the software and more time on the modeling. And so data science has come full circle – from an order of cluster-whisperers who could coax insights from data to mainstream professionals who can do whatever is needed (as far as math or coding) to solve the business problem.

2.1.3 Data Science Roadmap

The process of solving a data science problem is summarized in Figure 2.1, which I call the Data Science Roadmap. The first step is always to frame the problem: identify the business use case and craft a well-defined analytics problem (or problems) out of it. This is the most important stage of the process, because it determines whether the ultimate results are something that will be useful to the business. Asking the right question is also the place that managers and executives are most critical.

This is followed by a stage of grappling with the data and the real-world things that it describes. This involves the nitty-gritty of how the data is formatted, how to access it, and what exactly it is describing. This stage can also reveal fundamental inadequacies in the data that might require revising the problem we aim to solve.

Next comes "feature extraction," where we distill the data into meaningful numbers and labels that characterize the objects we are studying. If the raw data are text documents, for example, we might characterize them by their length, how often certain key words/phrases occur, and whether they were written by an employee of the organization. Extracted features can also be quite complex, such as the presence or absence of faces in an image (as guessed by a computer program).

If asking the right question is the most important part of data science, the second most important part is probably feature extraction. It is where the real-world and mathematics meet head-on, because we want features that faithfully capture business realities while also having good mathematical properties (robustness to outliers, absence of pathological edge cases, etc.). We will have much more to say about feature extraction as the book progresses.

Once features have all been extracted, the actual analysis can run the gamut from simple pie charts to Bayesian networks. A key point to note is that this loops back to framing the problem. Data science is a deeply iterative process, where we are constantly refining our questions in light of new discoveries. Of course in practice these stages can blur together, and it may be prudent to jump ahead depending on the circumstances, but the most important cycle is using our discoveries to help refine the questions that we are asking.

Finally, notice that there are two paths out of the workflow: presenting results and deploying code. These correspond to the case when our clients are humans making business decisions and when they are machines making judgments autonomously.

2.1.4 Demystifying the Terms: Data Science, Machine Learning, Statistics, and Business Intelligence

There are a number of fields that overlap with data science and the terminology can be quite confusing. So now that you have a better idea what data science is, let me sketch the blurry lines that exist between it and other disciplines.

2.1.4.1 Machine Learning

Machine learning (ML) is a collection of techniques whereby a computer can analyze a dataset and identify patterns in it, especially patterns that are relevant to making some kind of prediction. A prototypical example would be to take a lot of images that have been manually flagged as either containing a human face or not – this is called "labeled data." An ML algorithm could then be trained on this labeled data so that it can confidently identify whether future pictures contain faces. ML specialists tend to come from computer science backgrounds, and they are experts in the mathematical nuances of the various ML models. The models tend to work especially well in situations where there is a staggering amount of data available so that very subtle patterns can be found.

Data scientists make extensive use of ML models, but they typically use them as pre-packaged tools and don't worry much about their internal workings. Their concern is more about formulating the problems in the first place – what exactly is the thing we are trying to predict, and will it adequately address the business case? Data scientists also spend a lot of time cleaning known pathologies out of data and condensing it into a sensible format so that the ML models will have an easier time finding the relevant patterns.

2.1.4.2 Statistics

Statistics is similar to ML in that it's about mathematical techniques for finding and quantifying patterns in data. In another universe they could be the same discipline, but in our universe the disciplines evolved along very different paths. ML grew out of computer science, where computational power was a given and the datasets studied were typically very large. For that reason it was possible to study very complex patterns and thoroughly analyze not just whether they existed, but how useful they were. Statistics is an older discipline, having grown more out of studying agricultural and census data. Statistics often does not have the luxury of large datasets; if you're studying whether a drug works every additional data-point is literally a human life, so statisticians go to great pains to extract as much meaning as possible out of very sparse data. Forget the highly tuned models and fancy performance specs of ML – statisticians often work in situations where it is unclear whether a pattern even exists at all. They spend a great deal of time trying to distinguish between bona fide discoveries and simple coincidences.

Statistics often serves a complementary role to data science in modern business. Statistics is especially used for conducting carefully controlled experiments, testing very specific hypotheses, and drawing conclusions about causality. Data science is more about wading through reams of available data to generate those hypotheses in the first place. For example, you might use data science to identify which feature on a website is the best predictor of people making a purchase, then design an alternative layout of the site that increases the prominence of that feature, and run an AB test. Statistics would be used to determine whether the new layout truly drives more purchases and to assess how large the experiment has to be in order to achieve a given level of confidence.

In practice data scientists usually work with problems and datasets that are more suitable for ML than statistics. They do however need to know the basic concepts, and they need to understand the circumstances when the careful hair-splitting of statistics if required.

2.1.4.3 Business Intelligence

The term "business intelligence" (BI) focuses more on the uses to which data is being put, rather than the technical aspects of how it is stored, formatted, and analyzed. It is all about giving subjective insight into the business process so that decision makers can monitor things and make the right call. As a job function, BI analysts tend to use data that is readily available in a database, and the analyses are mathematically simple (pie charts, plots of something over time, etc.). Their focus is on connecting the data back to the business reality and giving graphical insight into all aspects of a business process. A BI analyst is the person who will put together a dashboard that condenses operations into a few key charts and then allows for drilling down into the details when the charts show something interesting.

This description might sound an awful lot like data science – graphical exploration, asking the right business question, and so on. And in truth data scientists often provide many of the same deliverables as a BI analyst. The difference is that BI analysts generally lack the technical skills to ask questions that can't be answered with a database query (if you don't know what that means, it will be covered later in the book); they usually don't have strong coding skills and do not know techniques that are mathematically sophisticated. On the other hand they use tools like Tableau to produce much more compelling visualizations, and typically they have a more thorough understanding of the business.

I mentioned earlier that in data science sometimes the "customer" is a machine that will run an algorithm autonomously and other times it is a human who will make a decision. This is another way to break down the disciplines. ML experts usually have machines as their customers; they are producing sophisticated models that will be used to power software applications. Statisticians and BI analysts are producing results for human consumption.

2.1.5 What Data Scientists Don't (Necessarily) Do

A common problem is that people have an overly broad definition of the field: data scientists are general-purpose geniuses who can do anything reminiscent of math, and you can give them any problem. This misconception is partly because data science is inherently a "jack of all trades" discipline and partly because some data scientists (especially in the founding generations) *are* geniuses and have a strong background in many related areas.

I would like clarify the situation by outlining a few things that do not fall under the umbrella of data science. A given data scientist might be able to do them, but it is not standard-issue.

2.1.5.1 Working Without Data

Paradoxically, the single most common inappropriate request I have seen is to do data science without data. This seemingly paradoxical request is actually forgivable, because quality datasets are often very difficult to come by. Say you are building an industrial machine and want to build a system to monitor it and predict failures, but the data isn't available because the machine doesn't exist yet. Or you want to understand the demographics of who your customers are, but you've never gathered that information. These situations aren't mistakes on the part of business leaders – they are actually side effects of being proactive – but they complicate things.

If you find yourself in this situation, you could use a data scientist to help set up the data collection system. However, the technical aspects of that would probably be better done by a data engineer, perhaps working in concert with a domain expert to understand what data are most important to store. There are two main use cases where a data scientist *might* be useful:

1. If there is a dataset that is related to what you would like to study, the data scientist can study it in the hope that any findings will generalize. However, you must bear in mind that:
   - Your findings are likely not to generalize. Gauging this risk is a question for domain experts, not data scientists.
   - You might hope to develop your data processing pipeline using the available dataset, in the hope that you can just drop the real data into it later. However, the dataset you have is likely in a different format, with its own idiosyncrasies, in which case it may need to be completely re-tooled when the real data becomes available. In that case you may as well have just waited.
2. You can simulate data, based on a theoretical understanding of what it *should* look like. This simulated data most emphatically will not give you any insights into the world, but it may be suitable for stress-testing your data storage framework or for training version 1 of an ML model.

I once had a client once who hired me to develop algorithms that recognized certain characteristic shapes in data plots coming off of machines they were building. There was no data because they had not yet deployed their machines, but there were some textbook pictures of what the plotted shapes should look like in theory. I wrote a script that generated hundreds of variations of each shape to see whether my algorithm could distinguish between them (billing at a high hourly rate), and it worked impressively well. However, when we finally got our hands on real data, it looked nothing like the textbook examples, and the abnormalities it contained were quite unlike the ones I had simulated. My algorithms failed on the real data, and all of the work I had done was ultimately useless.

Generally speaking, no data means no data science.

2.1.5.2 Working with Data that Can't Be Interpreted

Oftentimes you are able to get your hands on a dataset but making sense of it is another matter entirely. It is very common for a data scientist to be given a large dataset with no documentation, and then be forced to guess what it all means. This is especially the case in large corporations where different teams might have engineered different collection processes, there is legacy data, teams might not want to help each other, and so on. Industrial datasets are rife with edge cases that are reasonable in light of how the data was gathered, but that makes no sense without background knowledge. It is often not possible, given only the data, to figure out what the heck it means.

The opacity of these large datasets is not news, and there are companies selling software that they claim can automatically make sense of all the edge cases and provide a human-friendly interface to the underlying data. I am deeply skeptical of such claims, because even an intelligent human often can't make sense of all of the idiosyncrasies with confidence. In the real world you have to figure out what your data means; there is no shortcut.

The biggest problem with data you don't understand is that it hamstrings any attempt at feature extraction. Rather than carefully quantifying some aspect of the real world, you are forced to basically throw formulas at the wall and see what sticks. Chances are a lot of garbage will stick by dumb luck, and you won't actually get around to testing the formulas that work best. If somehow you do manage to find compelling patterns, it is then impossible to gauge whether they will generalize to slightly different situations in the future.

I had a situation like this at one point, joining together datasets from many stages in a manufacturing pipeline to try and predict failures in the final product. There were dozens of tables, with hundreds of variables, and barely anything was human-intelligible. It wasn't politically feasible to get an accounting of what the fields and tables meant, so I was forced to look blindly for correlations in the data. And indeed there were many variables that correlated strongly with failure, but every single one turned out (upon laborious digging around to figure out what those particular variables meant) to be something trivial that the engineers already knew well.

If you have a lot of data but don't yet know what it means, then hiring a data scientist or data science consultant is not yet worth your money. You should instead leverage domain experts (ideally one who were involved in collecting the data in the first place) to figure out and document what the different fields mean.

2.1.5.3 Replacing Subject Matter Experts

One of the most destructive abuses of data science is when people try to use it to replace domain expertise. Under the "working without data" section, I talked about how domain experts are critical for knowing what patterns to anticipate in a dataset, if you don't yet have the data itself. But they are also crucial for interpreting the data after the fact, and for extracting good features from it that can be used in predictive models.

The myth of replacing domain experts comes from people putting too much faith in the power of ML to find patterns in the data. As we will see in later chapters, ML looks for patterns that are generally pretty crude – the power comes from the sheer scale at which they can operate. If the important patterns in the data are not sufficiently crude then ML will not be able to ferret them out. The most powerful classes of models, like deep learning, can sometimes learn good-enough proxies for the real patterns, but that requires more training data than is usually available and yields complicated models that are hard to understand and impossible to debug. It's much easier to just ask somebody who knows the domain!

Data scientists may be able to wing it in solving everyday problems that many people are familiar with (measuring website traffic), or in situations where the raw data contains very few fields and it makes sense to use them directly as features. In more complicated situations though, a data scientist will have to work closely with domain experts to craft features that capture to reality on the ground. Or they can be the domain experts themselves, either by doing their homework on the field or starting out in it (see the upcoming discussion of "citizen data scientists" – learning a little bit of data science can go a long way if you're already an expert on a subject!). Ultimately data science is not a way to replace subject matter expertise, but to augment it with tools that allow for better data analysis and the creation of useful models.

2.1.5.4 Designing Mathematical Algorithms

This one might seem surprising, so let me explain. A lot of people need a mathematically sophisticated algorithm to solve some business problem, and so they hire a "math wizard" data scientist to develop it for them. The problem is that data science only covers a small subset of math, specifically techniques for uncovering patterns in a dataset. There is more to the world.

For example, let's say you run a delivery company and you want to figure out which parcels should go with which drivers and which routes they should take. You know how far it is from anywhere to anywhere, plus how many parcels each car can hold. You want to divvy the parcels up and plan the routes so as to minimize the total number of miles driven between all your drivers. This is a complicated task that may be exceptionally difficult, but technically it is a straightforward math problem with a single correct answer (or perhaps several answers that each take up the same number of miles). You may need a lot of brains and computational power to solve it, but you don't need any insights from historical data.

Many of these problems fall under the umbrella of "operations research," a grab-bag discipline that uses computational math to solve various logistics problems. It is a fascinating subject in its own right, and I would argue that its practitioners deserve the "math wizard" reputation more so than data scientists. ML guesses an answer based on the patterns it has seen previously. Operations research deduces the correct answer based on first principles.

For the most part ML and operations research solve different problems. Complicated logistics puzzles are not susceptible to brute-force pattern matching, and conversely ML can pick up on subtle patterns that no human would have thought to encode. There are gray areas, and operations research will often use data science to solve sub-problems, but they are different disciplines.

2.2 Data Science in an Organization

This section will move beyond the work of an individual data scientist and discuss how they function on a team and add value to an organization.

2.2.1 Types of Value Added

You know the kinds of problems data scientists solve, but I would like to enumerate some of the ways those problem can actually add value.

2.2.1.1 Business Insights

The most basic function of a data scientist is to provide business insights. Often these insights come as relatively simple one-off tasks, the results of which can be boiled down into a PowerPoint slide or even a single number. Some of the most bang-for-my-buck I've gotten is simply to help people formulate a SQL query that actually captures what they want to compute. Other times the questions are extremely open-ended, requiring weeks of iteration where different metrics are tried out (what counts as a "loyal user"? What sources of noise are present in this data again?) before the answers come, and they must be interpreted very carefully.

A critical distinction to understand is whether an insight is just nice to know, or whether it will be an ingredient in a business decision. There are infinitely many questions and variations of them that you can ask, and gain nothing from them except a psychological assurance that you're "well informed." It is better to use data science to test the assumptions underlying business decisions, identify pain points in an operation, and help make crucial decisions.

2.2.1.2 Intelligent Products

Many of the most high-profile applications of data science involve embedding logic into production software. This ranges from the famed Netflix recommendation engine to systems that identify faulty components on a factory line. The core of a system like this is typically a large-scale ML algorithm. The role of the data scientist can cover all parts of a system like this; defining the specific problem to be solved (including success metrics that capture business value), extracting the key features, crafting the mathematical model itself, and writing efficient code that can run in production. Something like the Netflix engine will, in practice, be worked on by a large team that includes many specialists in various subjects, but smaller projects will often be done by a one-person shop.

Usually though it's not possible for a single data scientist to do the whole project themselves; they write the code for the core logic, which is then integrated with the work of software engineers into the final product. Data scientists often don't have the specific technical skills required for the production code, which might be in an unfamiliar programming language and use software tools that are alien to them. A production codebase can be extremely complex, and it may be impractical for the data scientists to know its ins and outs in addition to their other responsibilities. In some cases the data scientist might simply not be a good enough programmer to make the code efficient. In any of these situations, the key question is this: where and how do you draw the line between the mathematical modeling and the production code written by software engineers?

I will discuss some of the possibilities in more detail later in the book, but I will start by discussing a common choice that rarely turns out well: having the data scientists tell the engineers what logic to implement. There is tremendous overhead required to communicate the algorithm in the first place, after which you have the daunting task of ensuring that two parallel codebases perform the same even as bugs are fixed and features are added. Usually it is better to figure out a way for the data scientists to write the logical code themselves and then plug it into a larger framework that is maintained by the engineers.

2.2.1.3 Building Analytics Frameworks

Data scientists are often the first people on the scene when a team tries to get organized with their analytics efforts, and there is little more than a bunch of data files when they arrive. While the main task of a data scientist is to formulate and answer questions, the first order of business in these cases is to get the data organized, placed into reasonable databases, and so on. Theoretically this is more the work of a data engineer – a software engineer who specializes in data processing systems – but it often falls to the jack-of-all-trades data scientist as a way to enable their own work.

After that backend is all in place, you're a large part of the way to a self-service system that can be made accessible to the rest of the team. This might include live dashboards, easy-to-use databases, and so on. Many data scientists create tools like these that democratize access to the data and automate mundane parts of their workflow. This lets the team get answers to common questions more quickly and allows the data scientist to spend their time on the trickier modeling questions.

2.2.1.4 Offline Batch Analytics

A midpoint between having data scientists write real-time production code and having them only provide numbers and graphs is the creation of batch analytics processes that run offline. Think of a process that runs nightly or at irregular intervals, and for which it's ok to take a long time (on the order of hours) to finish. The software engineering bar is not as high for batch processes as it is for a real-time product; efficiency is not a major or concern if multi-hour runtimes are acceptable, and if something goes wrong you just have an internal issue to resolve rather than something that is customer-facing.

Sometimes batch processes are used for reporting purposes, generating updated dashboards and metrics, and flagging any irregularities for human inspection. This is typical of the "clients are humans" approach. In other cases the batch process will tune the parameters of an ML model to the latest data, or populate a database that will be accessed in real-time by the production system.

2.2.2 One-Person Shops and Data Science Teams

Especially in small-to-medium organizations, it is common for a team to have multiple software engineers, managers, and the like, but only a single data scientist. There are several reasons for this, including the following:

- Many of the projects a data scientist takes on are small enough in scale for a single person to handle. Correctly formulating and answering a question has fewer moving parts than most pieces of production software.
- In creating a software product, the bulk of the work is usually in writing the production software, not in analyzing data. As such you often should have a large ratio of engineers to data scientists.
- Data science work can be cyclical. There may be a lot of work leading up to major business decisions, or upfront when key ML models are being developed, followed by a comparative lull.

On some teams there is usually only enough work for a part-time data scientist. In those cases it is common to have the data scientist devote their extra cycles toward some other task, like software engineering. Situations like this can also lend themselves to "citizen data scientists," employees who are primarily subject matter experts but have picked up the skills of a data scientist on the side and can step into that roll when needed.

As an organization gets bigger, it starts to make sense to have large-scale analytics infrastructure. In these cases there is work constantly going on to tweak the logic of running software, organize and generate regular reports, conduct large-scale tests, and be ready to handle one-off questions about any aspect of the organization.

2.2.3 Related Job Roles

Data scientists are often not the only players in the modern data ecosystem. This section will review some of the other job roles as they relate to data science.

2.2.3.1 Data Engineer

A data engineer is a software engineer who specializes in designing, developing, and maintaining data storage and processing pipelines. These are the kinds of tasks that a data scientist might do in a hacky way, but that will be done in a much more robust way by a data engineer. A data engineer can be compared with a web developer; each is a software engineer working in a highly specialized niche, with a large corpus of technologies and best practices that a normal software engineer is unlikely to know.

In large-scale data operations, it will be the data engineers who set up the relevant databases, data ingestion systems, and pre-processing pipelines. In short, they create the environment where data scientists can focus on the content of the data rather than IT concerns. They typically will not know much about statistics or ML, but they work with data scientists to ensure that the systems will be extremely performant under expected use cases. Standard data processing tools – like relational databases and Big Data pipelines – are liable to be done by either a data scientist or a data engineer (though the latter will typically do a better job). When you start talking about parallelized ways to load data into those tools, or making them real-time responsive, you are solidly in the data engineer's wheelhouse.

Some people who think they need a data scientist would do better to hire a data engineer. If you have a daunting pile of messy data, but your actual questions are simple and precise, they are likely to do a substantially better job (and possibly for a somewhat lower salary). Table 2.2 shows some typical jobs that would be more appropriate for one role versus the other, or which are a good fit for either.

2.2.3.2 Data Analyst

As I discussed previously, analysts specialize in creating business insights through mathematically simple operations on data, especially the creation of charts and summary statistics. Those charts and statistics will generally be made with the help of a software framework – likely a well-known product such as Excel or Tableau, but possibly an internal tool that was created by a data scientist/engineer.

Analysts will often have a better understanding of the range of datasets available and of organization as a whole, including how the edge cases in the data

Table 2.2 Data engineers specialize in creating software systems to store and process large amounts of data, whereas data scientists focus on using those systems the create models and glean insights.

Data engineer	Either	Data scientist
• Creating a system to load data into a database in a parallelized way • Setting up a database so as to make certain known queries run more quickly • Writing a batch job that cleans known pathologies out of a dataset	• Getting a pile of CSV or Excel files into a simple SQL server • Calculating summary statistics from a dataset	• Any kind of machine learning modeling • Examining available data to craft good criteria for things like a "loyal user" or a "highly active session"

correspond to idiosyncrasies of business processes. As such they are an invaluable resource for data scientists, who are likely to "deep dive" on modeling a particular dataset and to have a litany of questions about it.

Especially when you are talking about domain experts who learn data analysis on the job, the distinction between data analyst and data scientist can get blurry. If somebody is tuning ML models, then he/she is clearly a data scientist, whereas if he/she works exclusively in Tableau, he/she is a analyst. But edge cases abound, like Excel power-users. Personally I generally draw the line at programming; if your analytics work includes a meaningful amount of programming in a coding or scripting language, then you are a data scientist, but if you stick to database queries, you are an analyst.

Table 2.3 shows some typical jobs that would be more appropriate for one role versus the other, or which are a good fit for either.

2.2.3.3 Software Engineer

Also known as software developers or "devs," these are the main people responsible for writing the code that goes into software products, either as a product that gets sold or as internal tools that need to be extremely robust.

Data scientists and software engineers use a lot of the same tools, so from the outside they can look quite similar. A key difference lies in the bar placed on the software in terms of code quality, documentation, reliability, adherence to best practices, and so on. A piece of production software is usually a large body of code, maintained and improved upon over a long period of time by a team and running in situations where it must be robust. This places a lot of additional requirements on production code and the way in which it is written, like

Table 2.3 BI analysts generally lack the ability to create mathematically complicated models or to write their own code (except possibly for database queries). However, they have deep knowledge of the business itself and are experts in communicating results in compelling ways.

Data analyst	Either	Data scientist
• Creating gorgeous charts for use in a press release • Putting together a dashboard of different charts to help monitor all aspects of a product	• Simple plots of business metrics over time to gain insights • Writing database queries	• Making predictions that go beyond fitting a trend line • Writing scripts to reformat data so that the graphing software can process it

- Teams and assignments change over time, so the code should be so thoroughly understandable that it can be used and modified without help from the original author
- Because the codebase is so large, it must be modular enough that one person can be maximally productive while only understanding a portion of it
- The codebase is expected to have a very long life, so the modules should be written in a way that supports likely future use cases
- Because the cost of failure is so high, there must be a system for thoroughly testing its performance and making sure errors are handled in an appropriate way
- Multiple people make changes to the same code, so they must be disciplined about coordinating their changes with each other

I don't mean to suggest that all production code does or even should follow all of these guidelines (they require a lot of overhead), but data science lies on the lax end of the spectrum.

Data scientists and devs tend to work together in two major areas; code that gathers or generates data and code that runs data science logic in production. When talking about code that generates data, the devs are often the from-the-horse's-mouth resource to understand what the data means, and to ferret out any issues contained in it. They don't just understand the data – they wrote the code that created it! They can also work with data scientists to ensure that the raw data is being generated in a format that captures all necessary information and is easier to process.

I already touched on the issue of data scientists working with devs when I discussed intelligent products. In some cases data science can be used to tune parameters in the production code or validate a simple piece of logic that will run in production. For more complex analytics though, like ML models or

Table 2.4 Software engineers create products of a scale and complexity far greater than data scientists can typically handle. However, data scientists often create logic that gets plugged into those products or analyze the data that they generate.

Software engineer	Either	Data scientist
• Creating a user-facing software product that needs to handle errors gracefully • Coordinating more than a handful of people working on a single codebase	• Creating an ugly internal website to help with annotating data • Writing the code that plugs a pre-trained machine learning model into a product	• Determining logical rules for how a piece of software should anticipate user actions • Analysis of the data generated by a software product

complicated regressions, it is usually a better idea to have the data scientists write the production code rather than having engineers create a parallel "production friendly" version of it.

The process of integrating data science code with the rest of the framework is often more frustrating than creating either piece in isolation. The solution is to have the data scientists and devs agree early on the ways that their code will interface, whether it is through a REST API, a batch process that populates a database, or a code module containing the logic. Then the data scientist can write their code, from the early stages, such that its input and output are in the agreed-upon format.

It's often a good idea to have the first version of the data science code be something that is logically trivial but that integrates correctly with the rest of the framework. That way you can make sure that the two codebases work together from the get-go. Otherwise you risk a last-minute disaster when they fail to integrate for some unforeseen reason and the whole system is non-functional.

Table 2.4 shows some typical jobs that would be more appropriate for one role versus the other, or which are a good fit for either.

2.3 Hiring Data Scientists

Hiring a data scientist is a large investment of time and money, even more so than other job roles because of their high salaries and the range of skills you'll have to assess them on.

This section will help you to assess your data science needs and hire the best people to fill them.

2.3.1 Do I Even Need Data Science?

Many managers aren't at the point where they have specific data science questions; they are facing a daunting pile of technical problems and they need somebody to help them sort it all out and do what needs to be done. The person for the job might be a data scientist, but they could also come from a host of other backgrounds. I have often been contacted about "data science" work when people really needed a mathematician, a data engineer, an analyst, or a programmer.

To help decide whether data science is right for you, I have compiled this checklist of general requirements:

- *You have data or can get it*
  Data science involves extracting patterns and insights from real-world data. If data isn't available you will need to figure out a way to get it. A data scientist might be able to help you with that, but it would be different from their core job function. If data isn't needed for what you want to do, then you are probably looking for a domain expert or a mathematician, who can solve the problem from first principles.
- *You know, or can figure out, how to interpret the data*
  Having the data as a bunch of files is not sufficient. You need to know what the various parts of it mean and how it is organized, or at least be able to figure it out. Datasets like web pages and written documents are typically not a problem, but some industrial datasets are opaque blobs of numbers and text, and it takes a domain expert to know what anything means.
- *You have open-ended questions OR you need an AI model*
  If you have questions that involve some ambiguity, like how best to quantify the loyalty of a customer, then a data scientist will be able to devise good ways to quantify this and vet how effective they are. If your goal is to develop an AI model that can make decisions autonomously, this also falls into the data science wheelhouse. On the other hand, if you know exactly what you want to compute, you may do better to just have a data engineer.
- *You expect to have more questions down the road, or to refine your original questions*
  Data scientists are expensive. If you only anticipate needing data science one time up front, for example, to create an ML model that you don't plan to update, you may want to hire a consultant instead.

If you meet all these criteria then congratulations – you could use a data scientist!

2.3.2 The Simplest Option: Citizen Data Scientists

The term "citizen data scientist" refers to people who do at least some data science work but for whom "data scientist" isn't their primary job title (at least until recently, in some cases). Typically they are domain experts who have picked up analytics skills as a way to make them better at their real job. In some cases they make the deliberate decision to learn skills that are unusual for their industry. Other times they come from a professional background (like engineering or the sciences) where a large fraction of people learn those skills by default. Depending on your needs, a citizen data scientist might be the right choice for you.

The biggest problem with a citizen data scientist is that they probably aren't proficient with more advanced techniques, both mathematical and computational. This can lead to grossly under-performant systems and long times-to-insight, especially if the solutions they generate rely on some of the more cutting-edge data science tools. The biggest advantage of a citizen data scientist is that they have an intimate knowledge of the domain. This expertise will give them a huge leg up in extracting meaningful features from complicated data. It will also be invaluable in the most important part of data science: asking the right questions.

Citizen data scientists often start off as power users of tools like Tableau, Excel, or data analysis packages that are popular in their industry. These tools do an excellent job of making common use cases easy while allowing advanced users to ramp up to higher levels. At some point though they all level off and have some edge case of functionality that is missing. At this point you typically need a scripting language like python. The prototypical citizen data scientist will spend most of their time as a power user of some conventional analysis software, resorting to scripting to fill in the gaps in functionality.

They are likely to set up databases, write scripts that clean and reformat data, automate simple processes, and generally use programming skills to grease the wheels of the "normal way" of doing things. Plus they are the in-house resource for more advanced mathematical or computational questions when they arise. What citizen data scientists don't typically do is set up sophisticated (in terms of scalability, fault tolerance, etc.) software frameworks, employ advanced techniques like deep learning, or write any production code.

I encourage you to pursue citizen data scientists, rather than dedicated professionals, if the following is true:

- You mostly need data science to extract human-understandable insights from data offline
- You do not anticipate any need for writing production software
- There is a lot of domain expertise required to make sense of your data
- You do not see why you would need the more advanced subjects like deep learning and Bayesian modeling

There are many ways for a normal employee to become a citizen data scientist. Honestly I am of the opinion that it's not nearly as hard as it's often made out to be. There are many books and training courses available that give a strong overview of the subject, and specific skillsets can be added piecemeal.

As better software tools are developed, it is becoming easier to use advanced analytics with a minimum of software engineering. This is especially the case with software packages that are specific to a particular industry or vertical, where it is possible to bake advanced techniques into very specific features, pre-tuning them so that they can be used without looking under the hood. Some of these products manage the data from initial collection all the way through to presenting analytics results, which reduces the need for the cleaning/reformatting that are often what drives people to scripting. This will lower the barrier to entry for advanced analytics, but it will also encourage people to ask increasingly complicated questions that will require resorting back to code.

2.3.3 The Harder Option: Dedicated Data Scientists

If you are looking to craft sophisticated predictive models, embed data-based logic into production code, or use advanced techniques like deep learning then you will probably need a dedicated data scientist. Most of the rest of this section will discuss specific skillsets they will need and how you can evaluate them. First though I would like to discuss the background to look for more generally.

Good data scientists usually have a strong STEM background, which gives them (and proves that they have!) the mathematical fluency to understand the advanced techniques. However, while this conceptual foundation is necessary it is not sufficient. A PhD physicist, for example, is liable to know almost nothing about machine learning or statistics. The ideal candidate will have a strong STEM background, plus some proof that they have the skills of a data scientist. The latter can include a lot of experience analyzing data, a data science certification, or even self-study. The techniques are not complicated, but very few people will have all of them unless they've acquired them deliberately.

2.3.4 Programming, Algorithmic Thinking, and Code Quality

Section 2.3.5 will give you a checklist of things to look for when hiring a data scientist, but one category deserves special discussion: programming. It is without a doubt the most important skill for a data scientist to have, forming an essential component of every stage of their workflow. A beginning data scientist who doesn't know math, or who has no knowledge about the industry in which they are working, can still get some useful work done while they work to fill the holes in their knowledge. But if they can't code they can't do anything.

Given this it may come as a surprise that peoples' coding skills are often shockingly poor. This is true for software engineers as well and is a well-known headache of the hiring process: a large fraction of people who have impressive STEM credentials are incompetent at programming. Hiring them can be a disastrous situation. At best they will simply get nothing completed – they will talk a lot and make a lot of suggestions, but there is never any work that you can put a pin in and move on. At worst they will patch together hopelessly over-complicated (near-)solutions to very simple problems, and force more competent programmers to either re-write the code or bend over backwards to accommodate their mistakes.

Many times in my consulting career, I have been a late-stage addition to an existing data science team, brought on in the hope of salvaging a project that seemed to be entering a death spiral as a deadline loomed. This has given me a lot of opportunity to see how things can go wrong. I have never seen a data science project fail for lack of sophisticated math, but I've seen many that fall victim to entry-level coding mistakes. A couple highlights that jump to mind are as follows:

- Another data scientist – a university professor with a focus on econ and ML – had quit the company right before I joined the team. It soon became apparent that a mission-critical script he wrote had a bug and I had to dive into his code. To my horror I found that almost all of his parameters were hard-coded, which made the script nearly impossible to modify and would have knee-capped the project if any of our constraints had changed. He also had deeply nested loops in his code rather than breaking them out into logical functions; the bug in his code was that in one place five tabs of indentation was visually indistinguishable from six tabs.
- A PhD physicist on the team had too much work to do before the deadline, and I had to finish one of the sub-problems that he had started. Not a single variable or function in his code had a name with a clear meaning, so it was impossible to untangle its logical flow. He explained the logic to me in English, and over a sleepless night I re-wrote the whole thing. My version was about a tenth as long as his, suggesting that the structure of his code was as convoluted as his naming conventions. I later found out that he had failed his coding interview and was hired on the basis of his academic credentials.

Sophisticated math and conceptual understanding are important to data science but they are not sufficient: the code also has to work.

The good news about basic coding skill is that it's relatively binary, so very simple litmus tests can weed most of the non-coders out of the hiring process. A common practice in interviewing coders of any stripe (which includes data scientists) is to open with a "fizz-buzz" question – a dirt simple coding question that will take a competent programmer a few seconds to solve. The classical fizz-buzz question is the following:

Write a computer program in any language that loops over the number 1 to 100 and prints them out. Except if the number is divisible by 3 print out "fizz" instead of the number, if it is divisible by 5 print "buzz," and if it is divisible by both print "fizz-buzz."

This should take a few seconds for a competent coder, but a large fraction of interviewees struggle over a long period of time and never manage to give a good solution. I always advocate giving interviewees a simple coding challenge (although maybe not quite this simple!) – and I always expect to be given one myself – even for very senior level jobs.

I think the central issue is what's called "algorithmic thinking," the ability to take a task that you know how to do and translate it into concrete steps. The steps do not need to be mathematically sophisticated – think of a fizz-buzz question – but they need to be clear enough in your head that you can express them in code. It's one thing to understand what the fizz-buzz question is and be able to write down the correct output yourself; it is another to write out the steps clearly enough that a computer can accomplish the task without understanding it.

A related issue is what's often termed "code quality," a collection of best-practices for writing code in a way that is organized, extensible, de-buggable, and understandable by another programmer. Algorithmic thinking is generally a prerequisite for good code quality, but is not sufficient. You learn to write good code by practicing it in situations where bad quality is punished (usually by negative feedback from other devs who need to work with your code), or by the trauma of having to work with somebody else's poorly-written code.

There is a virtuous cycle between algorithmic thinking and good code quality. If you can think clearly about an algorithm, you can code up the steps clearly and succinctly the first time, rather than jury-rigging a partial answer until it works. If the code you are working with is clearly written, then you can focus all your attention on the logical flow, rather than constantly trying to figure out what a variable means or why a step was taken. There is debate about where algorithmic thinking comes from in the first place; some people seem to be born with it while others get STEM degrees without ever becoming proficient. My experience has been that even if somebody doesn't have a natural flair for algorithmic thinking, good discipline about code quality combined with practice will be enough to do data science work.

Data scientists usually work with comparatively small codebases, and so their bar for code quality is not as high as for software engineers. There is still a bar though, and it matters: elementary lapses in code quality are often the root cause of data science disasters.

The best way to assess somebody's coding ability is to give them a hands-on technical interview or review their previous technical work. It will show up on a resume as a track record of writing their own code, especially in team environments. Educational credentials are not in themselves a reliable indicator of coding skill; classes focus on teaching a specific subject, and the projects are often short enough that they can get by with poor code quality.

2.3.5 Hiring Checklist

Here are the key things to look for in a data scientist, along with some specific criteria you can use to see whether they have them. Generally speaking, any of these criteria will give reasonable assurance that a data science candidate has this skill.

- Solid basic coding skills
  - o Doing well in a solid coding interview. In my mind this really is the gold standard, and I personally never hire a data scientist without one regardless of their credentials.
  - o Previous work experience as a software engineer at a reputable company. This could be full time work or repeated internships.
  - o Significant contributions to open source software projects.
- Basic knowledge of ML
  - o An academic background in ML, even just a single class. There isn't a huge amount of actual material to learn for an entry-level data scientist, and simple models are usually sufficient.
  - o A data science boot camp.
- Comfort with mathematics
  - o A degree in engineering, computer science, or one of the hard sciences. Many of these subjects won't teach the actual mathematics that data scientists typically use, but they can teach fluency with quantitative thinking.
  - o Experience doing significant amounts of statistics in another discipline, like biology or sociology. They should be able to go into some detail about the techniques they used and why they were relevant.
- Appreciation for simple approaches
  - o They can speak to examples of surprisingly simple analytics approaches.
  - o When approaching a modeling problem, they think of a simple approach first, even if they then reject it because it has some critical weakness.
- Experience working with real data
  - o Previous work as a professional data analyst
  - o Previous work as a data engineer

2.3.6 Data Science Salaries

Data scientists are notorious for commanding large salaries. This is partly because tech skills in general are well-paid, but also because they bring additional math skills to bear. Rigorous math and strong coding is a rare combination. Salaries vary from by region, city, and industry, but generally speaking a junior-level data scientist starts around $80 000 per year as of this writing.

I caution against trying to hire a self-styled data scientist for much below market rate, because you are liable to get what you pay for. Recall how many parts of an organization a data scientist interacts with, and how little oversight there often is of the technical aspects of their work. This makes a bad hire especially damaging.

If the salaries of a dedicated data scientist are too high for you, then you may want to consider a citizen data scientist.

2.3.7 Bad Hires and Red Flags

In this section I'd like to highlight several types of data scientist to avoid, and how they can be flagged during the hiring process. The groups are not by any means mutually exclusive, but they represent several dangerous trends:

- Mathletes
  Many data scientists have an almost compulsive need to apply heavy-handed mathematical tools for solving even simple problems. This is a problem because these techniques generally produce results that are harder to interpret, they often rely on idealized assumptions that are not true, and it is very easy to waste time fiddling with different variants of a fundamentally wrong approach. Simple techniques are easy to make sense of, easy to jury-rig, and they give result (or fail to) quickly. Additionally mathletes are often conspicuously lacking in the more basic skills like programming. Red flags for a mathlete include:
  - Excessive use of technical jargon
  - Difficulty explaining things in simple terms
  - An instinct to use complicated tools as a first approach.
- Academics
  Some people were not cut out for the real world. They are used to thinking very hard, over a prolonged period of time, about precisely formulated questions and giving rigorous solutions to them. These Academics fret and worry when idealized assumptions are violated, when steps can't be justified, and when there is pressure to move on to the next question before all the t's have been crossed. Ultimately data science is not about theorems or unassailable truth – it is about enabling good business decisions (even if those business decisions are made by a computer). There is a delicate line between being quick-and-dirty and being

reckless, but rough answers are often good enough. Academics strive to formulate problems that are solvable in a truly rigorous way, rather than working to put a fork in the current question and move on to the next iteration. Red flags for an academic include:

- ○ The only datasets they've ever worked with are well-studied academic ones.
- ○ They've only worked in academic settings, or heavily research-focused roles in government and industry.
- ○ They don't think about why a problem needs to be solved. A practical data scientist should be able to explain how solving a problem added value to their organization, and why simpler approaches wouldn't be good enough.
- ○ They can talk about different analytics approaches and their strengths and weaknesses, but they say very little about problems like data pathologies or code quality.

- Blowhards

To put it simply, a disproportionate number of data scientists are arrogant jerks who think they're smarter than everybody else. I'm not sure whether these people are drawn to the profession in the first place or whether the cache' of data science has gone to their head, but my experience is that they are actually less competent than average. These individuals will use the implicit authority of being a data scientist to impose their decisions on other people and insist that work be done in a way that accommodates their idiosyncrasies. The breadth of their knowledge (if not their wisdom) makes it difficult for other employees to argue with them, because they can bludgeon people with jargon from a range of disciplines. Managers without adequate technical background are sometimes taken in by their force of personality, giving their whims outsized authority. Because of the number of different areas that data science touches, this can be absolutely toxic to an organization. Red flags for a blowhard include:

- ○ They are hesitant to speak plainly about the limits of their knowledge and experience. Data science is too big a discipline for anyone person to be expert in all parts of it.
- ○ They are given to talking about how the problems they are working on are singularly difficult. They may do this in an attempt to make their work sound more impressive, or perhaps to explain away why they haven't made more progress in it. Note that simply pointing out that their problem requires a lot of leg work – like edge cases and sub-problems – is not in general a red flag.
- ○ They talk about the problems that other people are tackling as if they are trivially easy by comparison.

2.3.8 Advice with Data Science Consultants

Aside from hiring your own data scientist, a common choice is to simply have them consult for you. The advantages of this approach include:

- If you only need data science work sporadically – which happens a lot in small-to-medium sized organizations – you can only have them on the payroll during those times.
- They may be able to help in training some of your existing employees to be citizen data scientists. In my experience there is often a lot of low-hanging fruit in this area that you can pick by having the data scientist do pair-programming with your employees.
- If your problem requires some of the fancier data science tools, but you don't have enough work like that to justify hiring somebody full-time, then a consultant is the obvious choice.

The biggest disadvantage is that data science benefits immensely from having an intimate knowledge of the subject at hand. This will be lacking in somebody who only works with you for a short period of time, especially if they consult for clients across several different industries.

The data science consulting industry is an interesting place. There are a number of firms that specialize in it, but most of the consulting that gets done is in the context of another project. Many software companies – especially those offering cloud computing services or analytics products – provide some level of data science consulting to help clients get the most out of their main product. Early stage companies will often use extensive consulting as a stop-gap to pay the bills and develop a customer base while their main product comes up to snuff.

If you choose to go the consulting route I can give you the following advice to get the most out of the process:

- Make sure you have specific problems in mind that require a data scientist's skillset. See the section on "Do I Even Need Data Science" to assess this – examples would include tackling open-ended business questions or developing AI models that you will use. If your goal is mostly just to get your analytics house in order, rather than specific thorny problems, then you might do better to hire a data engineering consultant instead. Chances are most of the work that needs to be done consists of organizing the data, documenting what it all means, and making it accessible throughout your organization. This is work that is better suited to an engineer; and bear in mind that a good data engineer can do simple analytics himself if there is a need for it.
- Find a person in your organization who knows the available data and put them in contact with the consultant, making it clear that it's important for them to facilitate the data science work. High-bandwidth communication with this

point-of-contact will allow the consultant to come up to speed quickly on the data and its idiosyncrasies, so that they can quickly make sense of their results and come up with different approaches. Oftentimes nuts-and-bolts questions like this get relayed through various project managers; the goal is to control the messaging, but it often devolves into a noisy game of telephone. A lot of time gets saved if you put the technical individual contributors in direct contact.

- If the consultant is not already familiar with your discipline offer to arrange at least one meeting with a domain expert to discuss the sorts of patterns that are liable to occur and be relevant. Many of the most fruitful discussions of my career involved a domain expert describing a phenomenon in English and me brainstorming ways to quantify it in math.

- If they are providing code that will plug into your production system, hammer out the details of how they will integrate as quickly as possible. Ideally this should include:

  - Examples of the input and output data in the format that will be used in production, so that everybody can make their code work with that format. Make sure it's actual files: if you rely on descriptions (even written ones like in an email or a contract) you are virtually guaranteed to have an inconsistency. It can be something as simple as getting the capitalizations wrong or as significant as required information not being available.

  - An understanding of what software options are permissible. Oftentimes data scientists will develop their solutions using relatively obscure libraries that cannot be installed on a production machine. Sometimes the deliverable needs to be in a specific language like Scala or C++. And make sure that you pay attention to what versions of software are acceptable – python 2 and python 3 are not backward compatible, for example.

  - A description of how the software will be run and what constraints there are on it. Sometimes the production system has limited computational resources or is running an incompatible operating system.

Usually I recommend going a step further and having the data scientists provide a preliminary deliverable to your engineers that doesn't work well, but that has inputs and outputs in the format that will be needed in production. There are *always* hiccups when it comes time to integrate a model into production. These range from trivial data formatting issues to disaster scenarios, and they can cause a project to be delayed or derailed. The details of course are not set in stone, but you want to avoid last-minute surprises by integrating the end-to-end system early and re-testing it if the specs change.

2.4   Management Failure Cases

Finally in this chapter I would like to discuss several common failure cases in how data scientists can be managed. Data scientists are expensive to have on staff, so it's important to manage them in a way that gets the most bang for your buck, as well as encouraging retention. These failure cases fall into two general categories: giving them problems that are not reasonably solvable and solving problems that don't utilize their unique skills. They are all exaggerated versions of dynamics that are normal in the work of a data scientist, so it's important to monitor and make sure none of them are getting out of hand.

2.4.1   Using Them as Devs

It is common in business that there are no burning analytics questions at the moment but there is an urgent need for people who can write decent code. In these situations it is common for data scientists to temporarily take on the role of software engineers, usually making modules that can be plugged into the overall codebase by full-time devs. Especially in small organizations this is often the most pragmatic way for the company to utilize its employees, given their skills and the work that needs to be done. And I am of the opinion that a certain level of this is healthy for data scientists too, to keep their engineering skills sharp.

It should, however, be a temporary situation; after the storm passes the data scientist should go back to analytics-oriented duties. While working as devs they are generally going to be less effective than the full-time engineers (unless they ramp up fully on the production codebase, and it shouldn't get to that point) while commanding a higher salary. It is also likely to hurt their morale and increase the likelihood of turnover. If you think there's a good chance that you will run out of problems, then you may want to consider a data science consultant, or hire somebody who is explicitly willing to do a hybrid role.

2.4.2   Inadequate Data

Data scientists need data in order to get their work done. It needs to contain signal, there needs to be enough to solve the problem at hand, and there needs to be some understanding about what the data means. "Can a data scientist reasonably be expected to solve the problem with this data?" is an important question to ask.

The most common deviation I've seen from this is situations where people don't understand the data, and the data scientist is left guessing how to make sense of it. They spend their time guessing the meaning of different fields, ferreting out irregularities in the data, guessing how to accommodate them, and testing whether those guesses pass muster. This guesswork is unavoidable in some cases, but it is also unpleasant work that grossly underutilizes a data scientist's skills and will generally result in low-quality analytics.

Expectations about data quality can, however, go too far. I've seen a situation, for example, where the earlier data was in a different format from the later data and the data scientist complained that it was a waste of his valuable time to write a conversion script. Instead he suggested that all of the first round of data be re-taken by the lab techs (this was in an engineering research context)! The task of re-formatting data often falls to a data engineer if there is one on the team, but the data scientist is next in line. A good guiding principle is that the meaning and formatting of the data needs to be understandable. But if the meaning is clear, they should be able to handle any format.

2.4.3   Using Them as Graph Monkeys

Visual presentation of analytics results, including charts and dashboards, is part of the work of a data scientist. And in many situations it's important that they look good – not just informative – for important presentations.

But there is a limit. It is not cost-effective to have a data scientist spend a lot of time fiddling with charts and visualizations for primarily aesthetic reasons, and it's not the kind of work that utilizes their skillsets. Additionally the software products that create the most beautiful graphics (like Tableau) are often not the kinds of tools a data scientist would use, because they don't support complicated analyses.

The task of creating aesthetically beautiful visualizations and dashboards falls more naturally to BI analysts. Analysts are also more likely than data scientists to be good at the UI side of dashboarding – organizing the graphs in such a way that business stakeholders have an easy time navigating to the information that they need.

2.4.4   Nebulous Questions

The core job function of a data scientist is to take business questions and provide quantitative answers, making reasonable judgment calls about modeling details. It is not their job to identify the most important questions in the first place, although they can (like any employee) be part of that conversation.

It's kind of like software. The core job of a dev is to take a precise description of what the software should do (like wireframes) and turn it into working code, making reasonable judgment calls about how to implement it. Figuring out what the software should do is a process that can involve many people (designers, marketers, etc.), and it is typically a business person who makes the final judgment call. If you tell a software engineer – especially one who is early in their career or doesn't know your industry – to "make something cool," then it's your fault if their software is useless. In the same way it is the job of a business person to know what an organization's needs are, so that they can help data scientists hone in on the right questions to ask.

The rule of thumb is that you should be able to ask a data science question clearly in English; making the question mathematically precise and answering it is the job of the data scientist. "Which customer demographics make us the most money per person" is a question that the data scientist can really sink their teeth into. Asking "what do people do on our site" is not.

2.4.5 Laundry Lists of Questions Without Prioritization

This failure case comes from the same maxim as the previous one: it is the job of management, not of data scientists, to understand what is of business value and what isn't (though obviously data scientists, like any employee, can be expected to use common sense and be part of the discussion).

A problem that data scientists sometimes face is that they are given a long litany of questions to investigate with no prioritization. This often is the result of somebody (often a group of people) brain-storming things that they are curious about, without considering the fact that each question will require a lot of work to answer thoughtfully. Questions like this can provide a certain psychological assurance to the person asking, but they often amount to a lot of busywork for the data scientist.

Glossary

Algorithmic thinking   The ability to break a simple workflow down into basic steps that can be programmed into a computer.

Citizen data scientist   An employee whose primary job is not data science, but who has acquired that skillset and can apply it to their team's work. Often citizen data scientists are invaluable because of their ability to ask the right questions and make real-world sense of the answers. This often overshadows and weaknesses they might have in data science itself.

Code quality   A collection of best-practices in programming that make code easier to understand and work with. This includes things like how the code is organized, good naming conventions, and documentation.

Fizz-buzz question   Extremely simple programming questions that function as a litmus test for whether somebody the most basic competence in coding.

Data scientist   There is no precise meaning of this term, but "somebody who does analytically-oriented work that involves a lot of programming" does a pretty god job of capturing how it gets used in practice.

Data engineer   A software engineer who specializes in creating pipelines for storing and processing data.

Dev   A software engineer.

3

Working with Modern Data

This brief chapter gets to the heart of modern data science: the data itself. It will discuss ways that data can be generated, the format it is in, and the ways that it is stored and processed. The answers to these questions put hard constraints on what questions you can ask and make certain analyses overwhelmingly more practical than others. If business insight tells you what questions are important to ask, then it is the data that tells you which ones are possible to answer.

I used to buy into the fallacy that the only important thing about data was its information content: what it told you about the world. That's the whole point, isn't it? But when you deal with industrial-scale datasets, low-latency requirements for user-facing apps, or the flexibility to capture the complexity of human behavior, then you discover something else: the logistics of how data is organized and accessed are as important as the actual information content.

This chapter will begin with a general discussion of how data is gathered in the modern world, especially some of the recent developments that have motivated the growth of data science. I will blend discussions of the data with some of the real-world situations that it is describing. I will then shift gears toward considerations about how data of various types are stored and processed. This section will be somewhat superficial, with technical details and specific technologies relegated to later chapters. Instead I will focus on business applications and the general classes of technology they might dictate.

3.1   Unstructured Data and Passive Collection

Traditionally the data an analyst or mathematician worked with was quite a bit different from what it is today. In the first place it was structured. "Structured" in this context is a confusing jargon term that just means it was arranged into tables with rows and columns (I prefer the term "tabular"). The pieces of data themselves can be numbers, dates, or raw text (or occasionally more complicated

Data Science: The Executive Summary - A Technical Book for Non-Technical Professionals,
First Edition. Field Cady.

things), but the key is that they are arranged into a tabular format. So a spreadsheet will generally be structured data, but an image file or Word document will not.

Structured data lends itself to various types of mathematical analysis. For example, you can take the average of a column of numbers, or count the number of distinct entries in a column of raw text. This makes it possible to express a wide range of business analyses, across any structured dataset, using a relatively small set of basic operations and concepts. There are many software packages (especially relational databases, RDBs) that make these operations easy and efficient, so structured data is the traditional lingua franca of business analytics.

Data scientists (like everybody) prefer structured data when they can get it, because it is so much easier to deal with. But they are also forced to work with a wide range of unstructured data in various other formats. There are two reasons for this.

The first is that advances in software have simply made it feasible to analyze large datasets, and many of the most interesting datasets (like a collection of all the webpages on the Internet) are inherently unstructured.

The second reason that data scientists work with a lot of unstructured data is not about data processing, but about data collection. Traditionally most structured datasets were generated with some particular application in mind, and they are organized so that the expected analysis operations would be easy to think about and implement. You can look at this as front-loading the analytics effort; if you do a good job of figuring out what tables you need and what columns they should have then your downstream analyses will be a lot easier. In effect the dataset is pruned down to only the things that are likely to be relevant.

In recent years though it has become a lot easier to passively collect massive datasets with no particular analysis in mind. You can record everything every user clicks on a website, along with when they clicked it. Or you can gather all of the raw log files that a computer generates, creating a stream-of-consciousness account of everything it is working on. The goal is not to make any specific analysis easier to perform; it's to make all conceivable analyses possible. This generally means taking an everything-and-the-kitchen-sink approach to collecting data and then storing it in a format that is close to the format in which it was gathered. The assumption is that later, when you have a specific analysis in mind, you can write code that distills this raw data down into only what is relevant and organizes it into tables that are suitable for the questions you are asking.

3.2   Data Types and Sources

Every dataset is unique to the business problem that it is measuring. However, there are several (not mutually exclusive) types of data that deserve special discussion; they are common enough that some variant of them shows up in many places, but each requires a certain degree of specialized processing:

- User telemetry data
  An example I gave in the last section – recording every time a user clicks something on a website and what they clicked – is an example of user telemetry data: there are people in the wild who are doing something, and we passively gather a record of what is going on. Typically user telemetry consists of event logs. Each piece of data will include a timestamp, the type of event that occurred (which may be an action the user took or a measurement that was made, like their heartrate), who did it, and any additional information specific to that event type.

- Sensor measurements
  As the Internet of Things becomes larger and more ubiquitous, there is more and more data that is being gathered by physical sensors. These include temperature, electrical properties of a system, and biometrics like heart rate. Typically these are expressed as time series, where the same measurement is made over and over again at regular (or irregular) intervals. It is my view that physical sensors like this are a new frontier for data science, since they require signal processing techniques that are not widely used at present.

- Natural language
  Documents of prose represent another type of data that requires highly specialized techniques. Human language is notorious for its ambiguity.

3.3   Data Formats

Data generally takes (or can take) the form of a file stored on a computer. Beyond understanding the content of that data on a conceptual level, it is also important to understand, at the most basic level, what the file looks like. Especially if you dabble in data science yourself, you will find yourself deeply concerned with the nuts-and-bolts of the file formatting.

My aim here isn't to give an account of all possible file formats; there are many of them around, and generally speaking their details are not important from a business perspective. But it is important to understand the tradeoffs involved in using one format versus another, and what applications they make sense for. To that end I will start by discussing three of the most important prototypical data formats: CSV, JSON, and XML/HTML markup. All of these formats are easy to understand, flexible, and ubiquitous in all areas of IT and data science.

3.3.1   CSV Files

CSV files are the workhorse of data analysis, the prototypical format for structured data. "CSV" usually stands for "comma-separated value," but it really should be "character-separated value" since characters other than commas do get used.

It is quite simple. Recall that a structured dataset has the data arranged into rows and columns. A CSV file is a plain text file that has one line of text for each row in the table, and the columns within a given line are separated by commas (or other characters). A few details worth noting are the following:

- They may or may not include the names of the columns as a header line.
- If the columns of data are text, and that text includes commas, then it becomes ambiguous where the different columns start/end within a given line. This is typically resolved by putting quotes around the text that includes a comma, like this:

  Field, Cady, "data science, the executive summary"

- Generally the file will not explicitly state the data types for the various columns (integer, text, date, etc.). Some software packages that read CSV will automatically make a best guess at the data type, and others will not.

A table in an RDB is basically a CSV file that has been loaded into memory in a particular way. Similarly, many data analysis software packages provide objects called "data frames," which are really just CSV files that have been loaded into memory in a way that is particular to the software that loaded them.

Besides their simplicity, a major advantage of CSV files is that they make mathematical analyses computationally easier. If a column of a data frame is all integers, for example, they can be stored in RAM as a dense numerical array like you would use in C, and arithmetic operations can take place at blinding speeds.

3.3.2 JSON Files

The CSV format is very constrained. Every row in a file will have the same number of columns, the columns will have to be of a consistent type (most software will forbid you from having a decimal number in a column that is supposed to hold a date, for example), and there is no notion of hierarchy. This works great if you know a priori what data will need to be gathered, and it has the advantage of enabling all sort of efficient mathematical operations in software, but in many real-world applications we need a more flexible format to store data in. JSON fills this need.

JSON is probably my single favorite data format, for its dirt simplicity and flexibility. Unlike CSV it allows for nesting of data, mixing and matching types, and adding/removing data as the situation dictates. The best way to illustrate JSON is by showing you an example of a particular JSON object:

```
{
"firstName": "John",
"lastName": "Smith",
```

```
"isAlive": true,
"age": 25,
"address": {
  "streetAddress": "21 2nd Street",
  "city": "New York",
  "state": "NY"
  },
"children": ["alice", "joe",
  {"name": "alice",  "birth_order": 2}]
}
```

At the highest level this object is enclosed in curly braces, which makes it a "blob." The blob maps different strings (like "firstName" and "lastName") to other JSON objects. We can see that the first and last names are strings, the age is a number, and whether the person is alive is a Boolean. The address is itself another JSON blob, which maps strings to strings. The "children" field is an ordered list. The first two elements in that list are strings, but the final element is another blob.

To express this a little more formally, we say that every JSON object is either:

- An atomic type, like a number, a string, or a Boolean. "null" is also a valid piece of JSON.
- A blob enclosed in braces {}, which maps strings to other JSON objects. The string/object mappings are separated by commas and their order is not meaningful.
- An ordered list of JSON data structures, enclosed in brackets [].

In the example earlier I split a single blob across several lines for easy reading, but I could equally well have put it all on a single long line. They are both valid JSON.

If you have a collection of JSON blobs that all have the same fields in them, and that don't do any nesting, then you can think of those fields as the names of columns and you may as well have stored everything in a CSV file. And indeed, in many software tools it is one line of code to switch back-and-forth between a CSV-style data frame and a list of JSON blobs. The CSV file has the advantages of being smaller (since you list the column names only once at the top of the file, rather than repeating them in every blob) and easier to process in memory. JSON has the advantage of flexibility – different blobs might have additional fields that aren't usually relevant, and they can be nested.

The ability of JSON to contain lists is particularly worth mentioning in contrast to CSV, because those lists could be of arbitrary size. In the example earlier is it easy to make John Smith have an arbitrary number of children. If we wanted to represent this data in CSV, we would need two separate files – one for parents and one for children – like the following:

Parents	
firstName	lastName
John	Smith

Children	
Name	Parent
Alice	John Smith
Joe	John Smith

A representation like this is immediately getting us into some advanced concepts of RDBs like foreign keys. It is much easier to just have a JSON blob one of whose fields is a list.

Perhaps the biggest advantage of JSON over CSV is that it allows for different fields to be added and removed and that only causes a problem for a piece of software if it operates on those particular fields. If your data looks like the example earlier, and all of a sudden the process that generates it starts adding additional fields saying what a person's hair color is, nobody's code will break – they simply won't use that field. This flexibility means that JSON is the most standard format when computer processes are passing messages back-and-forth between them.

3.3.3 XML and HTML

HTML is the lingua franca of the Internet: it captures the text of a website, along with all of the metadata that specifies things like italicized font, backgrounds of varying colors, embedded images, and the like. The primary function of a web browser is to take an HTML file – which is ultimately just text – and render it as a webpage. HTML is a special case of XML, and they are both called "markup languages."

CSV format is useful when you know exactly what the data fields are in every record. JSON is technically arbitrarily flexible, but in practice it lends itself to situations where 90% of the time you know what the structure of a data object will be. JSON also has no concept of metadata about a particular blob (you could jury-rig it by having a field in the blob called "metadata," but nobody does that). Markup languages are designed for situations where the data is truly free-form, with no expectation of consistency from one record to another, and records are likely to be embellished with metadata.

An example of XML (specifically HTML) is this:

> *Here is some sample website text. <i>This is italicized</i> and <a href="http://www.google.com">this goes to google</a>.*

The text is broken up into "elements" that can have special properties, like being italicized. An element's beginning and end are denoted by tags in angle brackets, like <i> and </i> – the "/" character means that the element is ending. In the case of the Internet's HTML format, the "i" element means that text is italicized and the "a" element means that you can click on the text to follow a link. The href="http://www.google.com" is metadata that assigns a "property" to the element. A given element can have as many properties as desired (including none). As with JSON you can nest elements in XML arbitrarily deeply, as in

> *All of the text after <i>this point is italicized, but <a href="http://www.google.com">this goes to google</a></i>.*

Note though that, unlike JSON, there is not a top-level element. An XML file can really just be a body of normal prose, with occasional elements marking it up to denote unusual formatting like italics.

An XML document can, in general, have any types of elements and any properties associated with them. But most XML documents you will work with are from some domain-specific dialect of XML, like HTML, where there is only a finite set of element types that you can use and each type has a finite set of properties you can associate with it.

Another example of XML files is the SVG file format for cartoon images. The elements in an SVG file denote geometric object like circles, lines, and other shapes that combine to form a complete image. Their metadata gives things like their color.

XML was originally intended as a general-purpose file format for exchanging data across the Internet, but for simple applications it has lost of lot of ground to the much simpler JSON.

3.4 Databases

The simplest way to store data is simply as files on a computer, and honestly that's what I usually encourage people to do unless there's a good reason to use other systems (with backups in the cloud, of course). But this approach often doesn't work well. Any of the following situations could come up and render files unsuitable:

- There is too much data to fit onto one machine.
- You can't take the performance hit of reading the data off of the disk.

- You can't take the performance hit of sifting through all the data to find what you want.
- Multiple users will need access to this data.
- You want safeguards to ensure the data stays self-consistent as changes get made over time.

In all of these situations you will be forced to use a database. This means that databases lie at the heart of virtually any large-scale product or organization.

You can think of a database as being a piece of software for conveniently storing and operating on data that might have otherwise been stored in raw files. As we will see RDBs are basically equivalent to collections of CSV files, whereas document stores are the equivalent of JSON files. CSV and JSON files are dirt-simple and perfect for long-term storage or quick-and-dirty applications. Databases on the other hand are sophisticated software programs that make data access fast and scalable, and suitable for the needs of massive organizations. But the information content is the same[1].

A prototypical database is a program running on a burly server, which holds more data than would fit into a normal computer, stores it in a way that it can be quickly accessed (this usually involves a ton of under-the-hood optimizations that the database's users are blissfully ignorant of), and is standing at the ready to field requests from other computers to access or modify the data. The main advantage of a database relative to raw files is performance, especially if you are running a time-sensitive service (like a webpage). Databases also handle other overhead, like keeping multiple copies of data synced up and moving data between different storage media.

3.4.1   Relational Databases and Document Stores

For our purposes there are two main types of database that you should be aware of: RDBs and document stores.

RDBs store structured data in tables with rows and columns. That data can be accessed with "queries" that say which data is desired and perhaps some pre-processing or aggregation that is to be done. The most historically important RDB is SQL, and nearly every existing RDB uses a variation of SQL's original query language. Most databases are relational – if you hear somebody say "database" without qualification it is almost always an RDB, and for that reason this book will spend the majority of its time on them.

––––––––

1 This isn't entirely true – many databases include metadata that augments the core data. For example, RDBs often allow for the notion that one column in a table contains unique IDs for the rows, and that those IDs map to a similar column in another table.

While RDBs store rows of data in tables, document stores typically store JSON objects (although in some cases they use XML) in "collections." This makes them more flexible than RDBs in terms of the data that they can store, but you pay a price in that they have less ability to run complex queries and aggregations. At the extreme end some document stores only allow you the basic CRUD operations – creating, reading, updating, and deleting one document at a time – and there is no way to have the database aggregate multiple records.

Document stores often form the backend for large applications that are accessed by many users. Take a simple phone app, for example. You can store the entirety of a user's information within a single JSON blob – including the arbitrary-sized data fields that make JSON so flexible – and as the person uses the app the content of their JSON blob (but only theirs!) evolves. Typical document stores are the open-source MongoDB and the AWS service DynamoDB.

3.4.2 Database Operations

Databases have two main functions. Firstly they store data in a way that is scalable, accessible, and usable by multiple people are services and has low latency. Secondly, they allow processing and aggregation of the data. RDBs are especially powerful in the processing they can do, but many document stores offer a large portion of that functionality as well. Databases offer a tradeoff: in exchange for limiting yourself to the operations that the database supports, those operations will be done quickly and efficiently against even a very large dataset.

I will discuss technical details more in a later chapter, where I go over the SQL query language. But in brief, the following operations will be possible in almost any RDB:

- Pulling out all the rows from a table that meet a certain criteria, like the "age" column being great then 65.
- Finding the min, max, or average value of a column. This is typically called "aggregating" the column.
- Matching up the corresponding rows in two different tables, which is called a "join." For example, I might have one table that contains employee IDs and their compensation, and another that gives IDs and their names. I can combine those into a single table that gives an employee's ID, name, and compensation.
- Grouping the rows of a table by the value in some column, and computing aggregates separately for each group. This is often called a "group by" operation. For example, let's say we have a table of employees, which tells us their compensation and the state they live in. If we "group by" the state and take the average compensation for each group, then we will get the average compensation in each state.

- Finally these operations can all be nested, and that's where database queries become extremely powerful (if possibly a bit tricky to wrap your head around!). Within a single query you can use a "group by" to calculate the average salary of employees in each state, and then join that against a table of state information to see whether salary correlates with the state's cost of living.

By combining these operations you can do an impressive amount of analysis of a dataset, and many data scientists work mostly through complex database queries. I talked previously about "business intelligence" (BI) as a separate job role from data scientists – a rule of thumb is that all BI analyses can be done with RDB queries, plus some software for visualizing the output of those queries.

Database languages are designed for relatively simple operations on data, with an emphasis on making it easy to answer easy questions with a one-off query. At the same time though they are capable of doing very complicated operations that are difficult to hold in your head at once. In industrial settings you will sometimes see queries that run into the dozens or even hundreds of lines and are carefully written by database specialists.

You can get a surprising amount done with only database queries, but there are limitations. The following operations will typically be impossible in an RDB:

- Training or applying any kind of machine learning model.
- While-loops, where an operation is repeated until some criterion is met. The operations an RDB will do are determined by the query you write; those operations (and only those ones) will happen regardless of what is found in the data.

To cast it in the language of theoretical computer science, RDB languages are not "Turing complete" – there are operations that are possible to do with a computer in general, but that are impossible in principle for most RDBs[2].

In exchange for accepting those limitations, you can write queries easily and run them at blinding speed over massive amounts of data. If you do need to do things that are impossible for an RDB, those are often done downstream, but a normal computer that is operation on the output of a database.

3.5   Data Analytics Software Architectures

I mentioned previously that simply storing files on a computer is the best way to conduct data science, unless there is a compelling reason to do something else.

---

2 There are some databases, like T-SQL, that are technically Turing complete. But these tools are designed primarily for more normal databases operations, with the features that make them Turing complete available in the breach. When people need the power of a Turing-complete language, they will usually use a more traditional programming language.

There usually is a reason to do something else though! This section will review several other common situations, and explain when you might use them. This list is far from exhaustive – most large-scale analytics systems will be highly customized and could incorporate aspects of all these approaches.

The schemes I will describe, roughly in order of increasing complexity, are (i) shared storage, (ii) a shared RDB, (iii) a document store + RDB, and (iv) storage and parallel processing. The appropriate level of complexity will vary from project to project and company to company. Generally I advocate using the simplest setup that is reasonable – this section will give you an idea of when it makes sense to move up to the next level.

In this section I will refer to the "cloud" a lot, and I would like to briefly clarify my meaning. Usually the cloud refers to services that are accessed over the Internet, like Amazon Web Services (AWS), which provides the infrastructure from many cloud-based analytics services. Other times though the cloud refers to physical servers that are owned by a company and accessed over the local network. The tradeoffs I discuss here apply equally well to both situations.

3.5.1 Shared Storage

Anybody used to a modern work environment will be familiar with the idea of shared documents that are stored in the cloud. These range from Google Docs to internal Wikis, and they make it easy for people to have common access to the documents. These same tools are sometimes used by data scientists to store data, with the intention of it being copy-and-pasted when they need it. Spreadsheet programs, for example, are great for allowing non-technical professionals to access and/or modify the data that is being used in analyses – the shared document functions as the master copy, and data scientists can export it to a CSV file if they need to perform analyses that aren't supported by the spreadsheet.

However, unless there is a compelling need to have human-friendly GUI interfaces, shared storage for data scientists is generally expected to be accessed programmatically. Typical examples that you will see are as follows:

- Industrial cloud-storage options like Amazon's S3 or Microsoft's Azure Storage. These options scale beautifully to even exceptionally large datasets, they are easy to use in concert with the data processing services offered by those companies, and they are affordable (at least, much more so than the data processing services).
- Small-to-medium datasets often get stored in files that are checked into software version control repositories. This has the great advantage that data scientists can use the same version control system for both their datasets and the code that operates on them.
- Many firms have a server that hosts files internally.

3.5.2 Shared Relational Database

In this situation there is a single RDB that stores all relevant data and is typically accessed by multiple people.

Oftentimes this situation arises organically: the production system is already storing its data in an RDB, so the most natural way to do data science is to pull data out of that same RDB. Oftentimes simple questions can be answered using only the RDB itself. Other times (visualization, machine learning, etc.), the RDB is used to generate a smaller dataset – pulling in only the parts of the data that are required for the analysis at hand and possibly doing some pre-processing – which is then fed into other software packages.

The key thing about an RDB – rather than simply shared storage – is that it allows for easy ad-hoc analyses:

- Running and writing queries is fast. In particular it can take mere minutes (or even seconds) to go from posing the initial question to getting a numerical answer.
- This short feedback loop means that it is possible to experiment with different ways of asking the question, without having to invest the effort of planning out a larger investigation.
- A surprisingly large fraction of the simple questions you might ask can be answered with just a query.

Using an RDB naively will usually outperform other ways you might perform the same computation. But the benefits are massively amplified if you take the time to set the tables up carefully. A table can be set up so that, for example, the data is internally sorted by a date column and so it is blindingly fast to analyze only the data for a particular time period. An RDB allows for these optimizations to be built in by database specialists, so that the performance benefits can be enjoyed by all of the other data scientists (who are often painfully ignorant about how to milk performance out of an RDB).

RDBs involve a fair amount of computational expense, and some of them do not scale well to extremely large datasets, so it is common to only have some of the available data in the RDB at any time. For example, it is common to only store the last year's worth of data, with the assumption that very few analyses will truly need to go back further than that.

3.5.3 Document Store + Analytics RDB

As with shared RDBs, it is common for data scientists to work with document stores simply because that is what the production software uses. They offer more flexibility than an RDS when it comes to the structure of data, and in many cases they are faster for the basic CRUD operations. This comes at the expense of being slower at operations that cover large portions of a dataset. In many document stores there are important analysis operations that are simply impossible.

In these situations it is common to create an RDB that exists in parallel to the document store and contains a condensed version of the same data. Typically each document in the document store will become a single row in the RDB, and that row will summarize the most important aspects of the more complex document. Syncing the RDB to the document store is a batch operation that can be run periodically or as-needed. Data scientists will run their analyses against the RDB whenever possible, occasionally resorting to pulling select data out of the document store when something is not adequately summarized in the RDB.

3.5.4 Storage + Parallel Processing

Databases offer low latency for basic operations, along with the ability to scale to reasonably large (or extremely large, depending on the database) datasets. But sometimes you need to run a computation that is impractical or impossible in a database – doing operations that the databases were not designed to support – at scale over a large dataset. This brings us into the realm of "Big Data," where the dataset is distributed across a number of different computers, which coordinate on a single large job.

The most important thing to understand about Big Data is that you should not use it unless you really need to. There is a lot of hype surrounding Big Data at the moment, and a widespread belief that it can do fundamental things that a single computer cannot. In fact the opposite is true: there are many important operations that can be done with no trouble on a single computer, but that are difficult or impossible in a completely parallelized system. The only advantages offered by Big Data systems are scale and speed: they can operate on datasets that are too large to fit onto a single computer, and they can reduce the time a job takes by having the computers work in parallel. As individual computers become more powerful, there is less and less need for data scientists to resort to Big Data systems.

That said, cluster computing is an invaluable tool for organizations that routinely run computations over massive quantities of data. Typically the cluster will be a shared resource, with IT infrastructure in place that schedules different people's jobs and ensures that no one user is monopolizing resources. The most popular parallel processing option in modern data science is Spark, which has largely replaced its predecessor Hadoop. I will talk in more detail about these technologies in a future chapter.

It is worth noting that many databases and data storage systems – especially cloud-based ones like those provided by Amazon Web Services – do in fact run on a cluster of computers. I put these into a different category from Big Data for several reasons:

- They are carefully designed so that users are shielded from most of the normal problems associated with cluster computing. It "just works" as easily as a normal database.
- They accomplish that seamlessness by being constrained in the functionality they offer, like a normal database. Systems like Spark provide more flexibility, but at the cost of having to worry about wrangling a cluster.
- They typically use an SQL-like query language, again like a normal database. More general parallel processing frameworks generally use a fully-featured programming language like Python or Scala.

Glossary

Blob A JSON data structure that maps strings to pieces of data (which can be numbers strings, lists, or other blobs).

CRUD operations A minimalist functionality offered by a data storage system. This allows you to Create, Read, Update, or Delete individual data records. It does not allow for fancier operations like aggregations.

CSV file A text-based "comma-separated value" file that stores data in rows and columns.

Document store A database that stores documents – typically JSON or XML. Document stores are more flexible than RDBs, but typically slower in data processing.

JSON A popular text-based format for storing data. JSON allows for both lists and for key-value maps, and can have structures that are nested arbitrarily deeply.

Relational database (RDB) A database that stores data into tables with rows and columns. This is by far the most popular type of database.

Structured data Data that is arranged into rows and columns, like you might see in a CSV file or an RDB.

Telemetry data Data that tracks the actions of users. Frequently it will be a collection of time-stamped events that records relevant actions as they interact with an app or webpage.

Internet of things (IoT) The emerging network of smart devices that is generating a growing amount of data. In many cases this is data gathered by physical sensors (GPS on phones, heart rate on smartwatches, etc.), which is often quite different to analyze from what most data scientists are used to.

Unstructured data Data that is not structured. This includes JSON blobs, XML objects, image files, free-form text documents, and most other file formats.

4

Telling the Story, Summarizing Data

I like to say that I have never in my career seen a data science project fail because people use overly simplistic math, but I've seen many go up in flames because people used math that was too complicated. A major theme of this book is that you should be cautious about using fancy techniques when simple ones will do, and this chapter discusses some of the simplest ones out there: drawing graphics and summarizing datasets into a few numbers. These are the types of business intelligence and statistics that predate data science but are still an essential part of the toolkit.

In subsequent chapters we will discuss predictive analytical models and artificially intelligent systems. The strength of these tools lies in their ability to find and recognize patterns, and even to make decisions, without human intervention. While this ability is game-changing, it is no substitute for human intuition and judgment.

In contrast the techniques in this chapter are all about aiding human understanding. They let us describe the world, quantify it, and rigorously test our understanding. They come from a time before it was technologically possible to have artificially intelligent systems making decisions autonomously, and we had to rely on old-fashioned human understanding. Modern techniques get most of the press in modern data science. But conventional analytics is as important as it ever was – much more so in fact, because there is just so much more data in the world.

This chapter has two main parts. The first is about "descriptive analytics": intuitive ways to quantify and visually explore data, looking for patterns and forming hypotheses. This is where you'll find graphics, summary statistics like the average and median, and ways to investigate the relationship between two numbers. Essentially descriptive analytics lets you distill raw data down into narratives that you can hold in your head. The second section is much more mathematical, and largely optional. It covers statistics, which lets you test those narratives rigorously.

Data Science: The Executive Summary - A Technical Book for Non-Technical Professionals,
First Edition. Field Cady.

It also discusses some of the most important probability distributions, the intuition behind them, and the sorts of real-world situations they could be used to model.

First though I will begin with some general discussions about picking what to measure, the importance of visualizations and outliers, and the role of experimentation in determining causality.

4.1 Choosing What to Measure

Most of the analytics literature focuses on mathematical aspects and ignores the most critical aspect of analytics: what you're actually measuring. In some cases the metric you should use is obvious, but in other cases you will have to decide exactly how to quantify whatever real-world phenomenon you're trying to capture. Feature extraction often requires a lot of judgment calls.

There is a strong tendency, especially in group settings, for people to raise a litany of (often legitimate!) concerns with any specific metric that gets proposed. It's not that the metric is fundamentally broken, but there are corner cases that aren't fully addressed, or maybe a fudge factor that somebody wants to introduce. This can quickly spiral into a "metrics by committee" situation, where you're measuring something quite complicated that doesn't fully make sense to anybody.

For example, say that you are measuring how many users you have for a website. Should you ignore people who are not on the site very long, and what is the cut-off? Should you only count return users, because first-timers might just be trying you out? Should you only look at users who have a "meaningful interaction" with the site, and what would that look like? Should power-users get double-counted in some way, because you want to measure the value people get from the site rather than just warm bodies? These are not frivolous concerns, but addressing all of them in a single metric will make it impossible to interpret and ask detailed questions about.

Nine times out of ten the rule is this: *for making business decisions measure the simplest thing that adequately captures the real-world phenomenon you're trying to study*. There are several reasons for this:

- It's easier to understand their strengths and limitations. There is no perfect metric, so you may as well opt for "the devil you know" whose limitations are easy to understand. This way you can assess how reliable the metric is and identify situations where it might break down.
- Communicating your results becomes much easier.
- Simple metrics are easier to carry over to new or related situations with a minimum of work.
- You're less likely to have bugs in the code that computes your metric, or edge cases that are handled inappropriately without being noticed.

- You're allowed to have a couple simple, concrete metrics rather than a single comprehensive one.

There are three situations that are partial qualifications to this rule:

- I caution against having lots of edge cases in the definition of a metric, but really what I mean is that you don't want a lot of judgment calls baked in. In some cases there is a long list of edges cases that are dictated by business logic. For example, there might be a list of products that are deliberately excluded from sales metrics because they get sold via a different avenue. In these cases there is usually clear logic that guides these rules in a black-and-white way, so you aren't really detracting from understandability. Plus these edge cases are probably listed out and carefully maintained before you put them into your metric, so the chance of introducing errors is muted.
- Sometimes metrics have a clear and simple business meaning, but a lot of complexity is required to faithfully measure that meaning. A common example of this in financial markets is the "value at risk," or VaR, of an investment portfolio. VaR is meant to measure the short-term riskiness of a portfolio, and it has the following meaning: according to in-house mathematical models, there is only a 1% chance of losing more money than the VaR in a given day. VaR is a controversial metric, and even its advocates think it should be taken with a generous pinch of salt (it may be more useful for assessing the reliability of the in-house models, rather than the portfolio risks!), but you do see model-based business metrics like this sometimes.
- I advocate simple metrics because they are easy to dissect and reason about. But in a few rare cases, you want metrics that are deliberately somewhat opaque, so as to discourage counter-productive nit-picking. The S&P 500 stock market index comes to mind; most people know that it's a weighted combination of 500 publicly traded stocks, but few people understand the mathematical intricacies of how that weighting is performed and adjusted over time. The goal is really just to take the temperature of a very complicated situation and move on. After all, actual decisions are generally made by more closely examining individual stocks.

The other thing to consider in designing metrics is how much accuracy is really required. To revisit the question of website traffic, for example, there is the thorny question of how to weed out bots – do you want to weed out as many bots as possible, or is it good enough to just have confidence they are less than 5% of the traffic you're counting? Bearing in mind that there is no such thing as a perfect metric, you should ask how big your error bars actually are before you worry about making them smaller. If the metric is adequate, move on for now and check back on it periodically.

4.2 Outliers, Visualizations, and the Limits of Summary Statistics: A Picture Is Worth a Thousand Numbers

The human brain is designed to process pictures extremely well, since vision is our primary sense. However, it is not really designed to process numbers greater than about five. As such you should strive to understand data with graphics and pictures, mostly using numbers to verify that understanding. When numbers must play a central role it's best to use pictures in addition, to make sure the numbers aren't misleading us.

A famous illustration of the disconnection between numbers and graphs is "Anscombe's quartet," pictured in Figure 4.1. The data in these four graphs are clearly wildly different. The top two graphs are a noisy linear relationship and a nice clean non-linear one. The bottom left showcases the effects of outliers, and the bottom right resists the very notion of a trendline. But according to the usual suite of summary statistics for data, these datasets are nearly identical. They all have the same line of best fit (shown), the fits are equally good, and the correlation between x and y is the same.

Anscombe's quartet is of course designed to capitalize on the weaknesses of those metrics, and there is a lot of focus on "robust statistics" that are more resilient to some of Anscombe's pathologies. The real lesson though is that any single number paints a woefully incomplete picture of your data, and you should strive to understand it in richer, more intuitive visual terms whenever possible.

There are two specific problems that make Anscombe's data a poor fit for classical summary stats:

1. *Modeling assumptions*: Metrics like correlation and trendlines tacitly assume that there is a linear – or at least a monotonic – relationship between fields in your data. This assumption is basically true for the graphs on the left, but not for the graphs on the right. In the top one a beautiful non-linear curve gets interpreted as noise. In the bottom one a single aberrant datapoint causes the analysis to hallucinate a line that isn't really there. Even robust statistics like the median tacitly assume that the median value is "typical," which is often not the case.
2. *Outliers*: The bottom row of graphs shows that a small number of outliers can throw off a metric dramatically.

In some cases there are rigorous statistical ways to test whether data meets your underlying assumptions, like the assumption that it's normally distributed. The problem is that usually either

- There isn't enough data to run the tests with confidence, or.

Anscombe's quartet

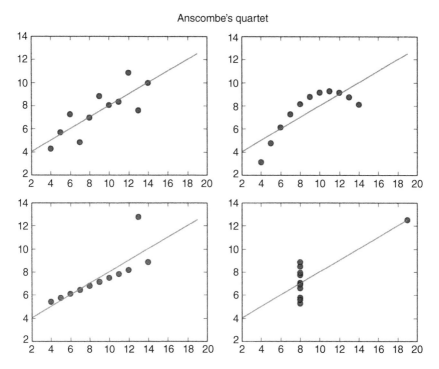

Figure 4.1 Anscombe's quartet is a famous demonstration of the limitations of summary statistics and the importance of visualizations. All four of these datasets have the same average and standard deviation for *x* and *y*, the same correlation, and the same trendline.

- There is enough data, and the tests fail because the assumptions are technically false. However, they're still approximately true and good enough to be useful.

There are various goodness-of-fit metrics that can partly work around the latter situation, but usually the best thing to do is simply to look at the data.

How to deal with outliers is a touchy subject that depends strongly on the problem you are solving. In some cases most of the outliers are garbage – the device taking the measurement failed, the website got hit by a denial-of-service attack, or something else happened that is outside the realm of the subject you're studying. In these cases outliers are often just discarded – before computing an average some people will go so far as to simply discard the top and bottom quarter of the data. In other situations though the outliers are the most important datapoints – the machines that are malfunctioning, the users who click on an ad, and the large transactions that really impact the bottom line.

What counts as an outlier is itself a little murky. Most of the theory of classical statistics was built up assuming that data follows a "normal distribution," the technical term for the bell-shaped curve in which large outliers are quite rare. In data that is normally distributed the outliers generally are rare, extreme, and usually safe to ignore because they represent some aberration in how the measurement was taken. In real situations though, there is often a "long tail" or "heavy tail" (I'll discuss more of this in Section 4.4) of gray-area datapoints, which are extreme enough to skew an analysis but common enough that you can't just ignore them. In these situations classical statistical metrics become largely meaningless, and we are forced to use "robust statistics" that are more resilient to outliers.

The importance of heavy tails can be illustrated by comparing two different demographic measures: people's weight and their net worth. The average weight is about 62 kg, and most people will be within a few tens of kilos from that average. Very few adults are less than half of the average or more than twice it, and those who are frequently have a medical condition that might justify excluding them from a particular analysis.

In contrast consider net worth. The median household net worth in America is about USD100 000, but fully 10% of households have a million dollars or more. The top 1% has USD8 million, or 80 times the average, and the disparity only goes up from there. For any reasonable definition of "rich," there is a significant proportion of rich people who are ultra-rich – this is the essence of heavy tails.

In my experience data science problems are usually like net worth. If you're analyzing human behavior, there will typically be a small group of power-users, which has its own core of ultra-users, and no clear cutoff where one group ends and the other begins. Popular websites receive vastly more traffic than unpopular ones, good clients bring in multiples of the revenue of bad ones, and so on.

4.3 Experiments, Correlation, and Causality

Usually the goal of data science is to find and characterize patterns in data, so that we can use them to make predictions in the future by assuming the pattern will continue to hold, and the goal of statistics is to carefully distinguish between real patterns and coincidences. Both of these are very different from the question of why a pattern is true in the first place.

Causality cannot strictly be determined by looking at passively gathered data, even if we can determine that a pattern exists. If people who shop on website A spend more money than those who shop on website B, it is possible that this is because A is better designed, but more likely it's just that the people who go to B are bigger spenders in general.

You've heard many times that "correlation does not mean causation." But the human brain has an almost pathological urge to read causal narratives in when two things are related in the data. General admonitions about correlation/causation are not enough keep us from leaping to conclusions – clearly *something* must be going on. As a more actionable alternative I suggest you keep the following in mind:

Rule of Thumb: *If X and Y are related in the data, then neither one is causing the other. Instead, there is some factor Z (called a "confounding factor" by statisticians) that is affecting them both.*

Descriptive analysis and visualizations can help us identify patterns in data, and statistics is used to help us make sure that the patterns we find are not just flukes. But the rule of thumb states that most of those patterns are not actionable. If we determine that X and Y are strongly correlated in the data, and then forcibly manipulate X in the hopes of influencing Y (say by re-designing our website to have the look and feel of the site where people spend more money), then chances are Y will remain unchanged and the correlation will simply cease to hold.

Instead we must figure out what Z might be (or hypothesize that in this case X *does* influence Y) and design a rigorous experiment to test whether it really works. The hypothetical causal factor is changed forcibly, with all other factors being held constant, and we see whether it has any effect.

An experiment is where we manipulate Z and look at whether Y changes in a way that is statistically significant. There are two key things you must keep in mind:

- Only Z should be manipulated. As much as possible everything else about the datapoints should either be the same or, at least, have the same variation between your test and control group. Usually this is done by randomly assigning points to one group or another – this is not always possible, but anything short of it leaves a lingering doubt that you have not found the true confounding factor Z.
- How much data is enough? To plan out how much data is required, you will need to know how big a pattern you hope to detect. If you are trying to detect biased coins, for example, a coin with 51% chance of heads will require an awful lot of flips before you can be sure there's a bias. Run an experiment long enough and there will almost always be some statistically significant difference between the test and control group. You need to decide how small a pattern it is important to detect, and be comfortable with the fact that you might miss anything that is smaller.

The logistics of running an experiment vary widely by industry. I have mostly done experiments in web traffic where we have a number of tremendous luxuries:

- There are many unique users generating different datapoints, and if you want to gather more data, you can just leave the experiment running.
- It is easy to randomly assign people to test or control groups.

- You find out within minutes whether a user took the action you are interested in.

Taken together this means that it is easy to iterate on different experiments we would like to run, getting results we can be confident in and then pivoting to the next tweak that we want to try out. Phone apps, software products, and websites are constantly using all of us as guinea pigs, because it's incredibly easy for them to do so.

In other industries it is not so easy. Datapoints are often expensive. It can take a long time to get results back, so you have to carefully plan out your experiments rather than iterating. And perhaps worst of all, manipulating only one variable may be impossible.

For example, let's say you are making pieces of consumer electronics and there are two different configurations of the components that you want to experiment with. The test and control group will in general require different tooling, with different skills needed to assemble them. So you may simply have to accept the fact that your test group was produced on one assembly line, by one team of people pulling components out of one bin, and your control group was produced by another. What if the configurations are equally good, but the people assembling the test components are better technicians? There's really no way to account for that possibility. Balancing the need for statistical rigor against the logistics of getting data can be a nightmare.

It is very rare to run a controlled experiment and then just "see what there is to see" as far as differences between the test and control groups. It is certainly fair (and a good idea) to do a variety of comparisons between the two groups and see if anything jumps out, but you run the very serious risk of making "discoveries" that are just statistical flukes. This risk becomes especially grave if you start asking fine-grained questions about small details that can vary between the test and control group. There are countless ways that they *could* be different, and if you look hard enough some of those differences will occur just by chance.

Instead you should go into an experiment with a plan about exactly what you will measure. This bit of planning is a good way to force yourself to clarify the business question you are trying to answer. More importantly though, deciding beforehand what you will measure keeps you from being led astray by spurious patterns.

4.4 Summarizing One Number

Many of the most illuminating moments in my career have been creating a histogram of data that, for one reason or another (usually just that there were so many fields to go over), I had not previously visualized even though I had been working with it for a while. On some level I always intuitively expect the histogram to be a bell-shaped curve, and so it's always exciting to see that it's richer and more complex than I had ever imagined. Often there will be a massive spike at some key number that drowns everything else out, and I have to filter out those datapoints before re-plotting. Other times there will be a long, thin tail stretching off to the right so that 99% of the data is scrunched over to the left – those large outliers must be removed too to make the picture understandable.

After those pathologies have been filtered and I finally see a proper histogram, it looks nothing like the textbooks. There may be multiple peaks, sharp cutoffs, and inexplicable slopes. Even if I don't know what it all means, I am clearly looking at the signature of a complicated, fascinating real-world process, rather than a sterile summary statistic. It is also humbling to realize that all this time I'd been treating it like a bell-shaped curve while the universe was snickering at my comically wrong assumption.

For that reason I encourage you to think of all one-dimensional data as really being a histogram, and the various summary statistics are (at best) crude ways to distill that histogram down into a single number.

That said, let's talk about some of the key properties that a bell curve can have.

4.5 Key Properties to Assess: Central Tendency, Spread, and Heavy Tails

4.5.1 Measuring Central Tendency

For purposes of understanding the real-world phenomenon that it's describing, the most basic question you could ask about a histogram is what its typical value would be. What counts as "normal"? This is a loaded question, and there are several common ways to answer it. These metrics are called "measures of central tendency," and the most common are the mean (aka average), median, and mode. They can have significantly different meanings, especially if the data is heavy-tailed; in that case you can visualize the differences in Figure 4.2.

For purposes of illustration let's say that you own a portfolio of websites, which cover a range of popularity levels. You are trying to understand the "typical" traffic that one of these websites receives.

4.5.1.1 Mean

The most common (especially historically) metric is the mean – the sum of all website visits divided by the number of sites. In mathematical circles it is conventionally described with the Greek letter μ ("mu").

The mean is not a measure of how much traffic is typical for a given site so much as a measure of how much traffic there is to go around in the entire portfolio. All of

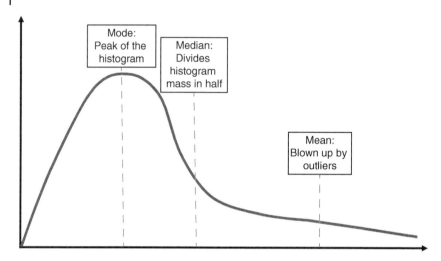

Figure 4.2 Mean, median, and mode are the most common measures of central tendency, i.e. trying to distill a "typical" value for a probability distribution into a single number. Each is frail in its own way.

the traffic could be going to one or a few sites, or it could be more equitably spread out among them – it makes no difference to the average.

For a bell-shaped curve without heavy tails, the average will be close to the point where the curve is highest. Plus it is quite easy to compute (especially by hand – averages as a tool long pre-date computers!) and has a number of nice mathematical properties, so it is very storied within the statistics literature. If you are working with data where you do not expect much in the way of heavy tails, then it is about the simplest metric you could hope for.

If you do expect heavy tails though, you should be aware that the average is very sensitive to them. In particular a heavy tail to the right of the histogram will make the average is far higher than most datapoints.

4.5.1.2 Median

Given how common heavy tails are in data, there is an increasing tendency to use the median as a measure of central tendency. The median is the middle value if you sort all of your data from smallest to largest – half the data falls below the median, and half above it. The median is a robust statistic because outliers have minimal effect on it; if the largest values in the data are extraordinarily large, it will not change the median.

For a perfectly symmetric bell curve, the median value will be at the peak of the histogram. A skew to the right will pull the median to the right, and it will end up in whatever place divides the mass of the histogram in half.

The median in probably the single most popular robust statistic and is probably the most common way to estimate what a "typical" value is for a numerical field.

4.5.1.3 Mode

The mode is the single most common value in the dataset, the point where the histogram peaks. In that sense it is maybe the most literal answer to the question of "what is a typical value." It also has the advantage that (unlike median and mean) it applies to non-numerical data – it's equivalent to asking which slice in a pie chart is largest.

The biggest problem with the mode is that data is often radically skewed toward one side of it or the other. In our example of website traffic, chances are the mode of this data will be zero even if only a few websites get no visitors, since it is unlikely that two high-traffic sites will have exactly the same number of users. So in this case the mode is not the "typical" website – it is an edge case whose importance is exaggerated by the fact that there are many different levels of traffic possible.

It would be possible to work around this by binning the sites into 0 visits, 1–100 visits, 101–200, and so on. But at this point you're opening a can of worms that kind of defeats the purpose – we want to have a quick summary statistic, not an investigations into the best binning system to use. If you do use a binning system, it is typical to have no cap on the highest-value bin, something like anything more than 10 000 visits to the website. Grouping them into a few bins underlines the key thing about using the mode: it works best when your data can take on only a few values, and you're looking for the most common one.

From this point on I will no longer discuss the mode; I will use the mean and median as the two measures of central tendency.

4.5.2 Measuring Spread

Aside from asking where the peak of a bell curve lies, the next natural question is how wide the bell is. Once we have established what a typical value looks like (whether the mean or the median), we want to know how close to it most of the data lies.

4.5.2.1 Standard Deviation

You may have heard something being described "several deviations above/below." This is a reference to the standard deviation, the classical way of measuring how spread out data is. You can think of the standard deviation as being the typical distance of a point from the mean, either above or below. In a classical normal distribution, 68% of data lie within one standard deviation, 95% within two, and so on. The standard deviation is usually denoted by the Greek letter σ ("sigma").

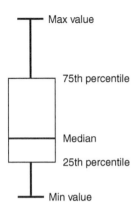

— Max value

— 75th percentile

— Median

— 25th percentile

— Min value

Figure 4.3 Box-and-whisker plots capture the median, the 25% and 75% percentiles (which can be used to gauge the spread), and the extremal values of a data field.

The thing to understand about the standard deviation is that, like the mean, it is sensitive to heavy tails and outliers. You can see this if you look at the mathematical definition. Let's say that you have numbers $x_1 - x_n$, and let's say that $\bar{x}$ is their average. The "deviation" of point x_i is $|x_i - \bar{x}|$, the absolute value of its distance from the mean; 0 is its exactly equal to the mean, and positive otherwise regardless of whether it's higher or lower. As the name suggests the standard deviation is sort of an average of these deviations, but with a twist: you square each of the deviations, take the average of those squares, and then σ is the square root of that average:

$$\sigma = \sqrt{\frac{\left|x_1 - \bar{x}\right|^2 + \ldots + \left|x_n - \bar{x}\right|^2}{n}}$$

Taking the squares of the deviations is where the sensitivity to outliers comes in. If the deviation is 5 then its square is 25, 10 then yields 100, and so on. If there is a single deviation that is much larger than the others then its square will dominate the average.

4.5.2.2 Percentiles

An alternative that is more robust to outliers is to use percentiles. It is common, instead of giving the standard deviation, to give the 25th and 75th percentiles of a dataset. This has the disadvantage that you have to use two numbers rather than one, and hence the convenience of using standard deviations as a unit of measurement. On the other hand using percentiles can more faithfully represent a situation where there is a long tail on one side of the distribution, because the 75th percentile may be much further from the median than the 25th percentile is.

In robust statistics it is common to summarize a distribution with five numbers – the min, max, 25th, and 75th percentiles, and the median – and to display them in what's called a "box and whisker plot" like Figure 4.3.

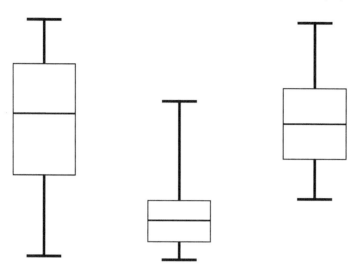

Figure 4.4 Box-and-whisker plots allow you to visually compare several data fields in a way that is quick and robust.

A box and whisker plot makes it very easy to compere several different distributions to get a sense of how similar they are, as you can see in Figure 4.4.

4.5.3 Advanced Material: Managing Heavy Tails

You actually hear people measuring the heaviness of tails whenever you turn on the news. Saying how much of the world's wealth is owned by the top 1% gives an indication of how lop-sided the distribution is.

Honestly I very rarely hear people putting a number to the heaviness of tails, and when I do it is usually what we just saw – stating that the top $X\%$ of the datapoints account for $Y\%$ of the total sum. The downside of these stats is that there is nothing magical about your choice of X or Y – 1% is mostly just a round number that sounds good. It is tacitly assumed that the top 1% is itself highly stratified.

A more mathematically principled way to measure the disparity of wealth (or correspondingly the heaviness of other tails) is the Gini coefficient. It ranges from 0 in the case of perfect equality to near 1 in the case that one datapoint has all of the mass. It has the nice property that whenever a dollar goes from a person with less money to a person with more, the Gini coefficient increases (even if both were already in the top 1%). Wealth in the United States has a Gini coefficient of about 0.4.

Heavy tails are one of the few areas where I feel like visualizations can be very misleading, and you should be wary of them. Would you have guessed from the left-hand graph in Figure 4.5 that the mean of one distribution is twice that of the other? At least to my eye it is hard to gauge just how fat a tail is – they all trail off to zero and are dwarfed by the rest of the histogram.

You can make heavy-tailed distributions more visualization-friendly by graphing not the raw numbers but the logarithms of them. The logarithm is a mathematical function that compresses large numbers down to a more manageable size. You can see the effect by comparing these two histograms, as in Figure 4.5. Increasing the logarithm by 1 means multiplying the number by a factor of 10, increasing by 2 means a factor of 100, and so on. Conversely, decreasing it by 1 means dividing by 10.

Besides reigning in heavy tails for visual ease, logarithms are often a better way to visualize data for the use case. Take the price of a share of stock, for instance. For investment purposes, going from USD2 to USD5 is the same thing as going from USD20 to USD50, and those differences will look the same if you plot the logarithm of the stock price rather than the raw price itself. Similarly, would you rather hear that somebody increased website traffic by 1000 visitors/day, or that they caused the traffic to double? When proportional differences are what you care about, then logarithms are the way to go.

One important nit-pick about logarithms is that they only apply to positive numbers. This is because repeatedly dividing a positive number by 10 makes it smaller and smaller, but it never actually reaches zero. Oftentimes the heavy-tailed distribution we want to examine is counting something, like visitors to a website, where zero is an entirely legitimate value. To deal with this it is common to look at $\log(n+1)$ rather than $\log(n)$; this will map zero to zero, and all larger numbers will be pulled down.

4.6 Summarizing Two Numbers: Correlations and Scatterplots

I've often joked that if humans could visualize things in more than three dimensions, then my entire job as a data scientist would consist of creating and interpreting scatterplots. In the same way that I encourage you to think of one-dimensional data in terms of histograms, scatterplots are the correct way to imagine two-dimensional data. As Anscombe's quartet showed, summary statistics can paint a misleading picture of the actual data.

4.6.1 Correlations

The word "correlation" gets used casually to indicate that two things are related to each other. It may be that one thing causes another: correlation between a cause and an effect may indicate a knob that you can turn to achieve a desired outcomes.

(a)

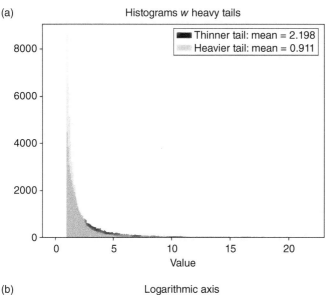

(b)

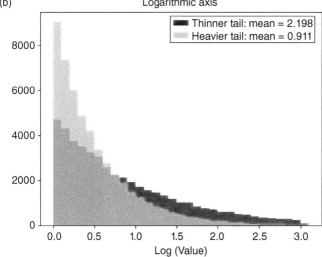

Figure 4.5 The histograms of two datasets, plotted for comparison on (a) a normal axis and (b) a logarithmic axis. In (a) the tails of the two datasets both trail off to being indiscernibly small. It's hard to gauge how many large datapoints there are in either dataset. But in (b) it is visually obvious that large outliers are dramatically more common in one dataset than the other.

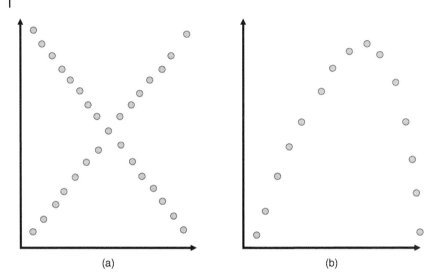

(a)   (b)

Figure 4.6 In both of these plots the correlation between *x* and *y* will be close to zero. Even robust metrics, like Pearson correlation, assume that there is a monotonic relationship between *x* and *y*.

If X and Y are correlated, it could be that a common underlying cause is nudging them both in the same direction (or opposite directions), or even that X and Y are two different proxies for the same thing.

Anytime there is a clear relationship between two variables in your data, it points to a potential insight about the situation you're studying. It is common, for example, to identify features that correlate strongly with a desired outcome and flag them for closer human inspection. Any such filtering though requires that we go from "there is a relationship" to a number we can compute.

The mathematical definition of "correlation" is quite limited in scope. There are several different ways to define correlation between X and Y, but they all measure a monotonic relationship where one goes up as the other goes up (or down) consistently. In both the scatterplots of Figure 4.6, for example, X and Y are clearly related to each other but the correlation between them will be near zero. We say that X and Y are "independent" if there truly is no relationship.

All correlation metrics range from -1 (when X goes up Y goes down) to 1 (X and Y go up together), with 0 being uncorrelated. There are three main types of correlation you are likely to see. The first is called the Pearson correlation – unless people specify otherwise this is usually what they mean. The numerical values of X and Y are used in computing it. The other two correlations are Spearman and Kendall. They are both "ordinal correlations" – they do not depend on the specific values of X and Y, but only on which ones are larger than which others.

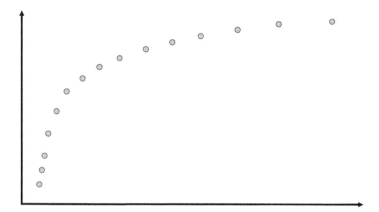

Figure 4.7 This dataset will have ordinal correlation of 1, since *y* consistently goes up as *x* increases. But it does *not* go up linearly, so the Pearson correlation will be less than 1. The Pearson correlation measures linearity, while ordinal correlations only measure monotonicity.

4.6.1.1 Pearson Correlation

Pearson correlation measures the degree to which your data lie on a straight line. Essentially you draw the line of best fit to the data; the size of the correlation indicates how good the fit is, and the sign of the correlation tells you whether the line goes up or down.

Pearson correlation is closely related to the mean and standard deviation. Not surprisingly then it shares their sensitivity to outliers and heavy tails. We saw in Anscombe's quartet in Figure 4.1 that a single outlier can badly throw off what would otherwise be the line of best fit.

In situations where you do not anticipate outliers (or where they are impossible – for example, if a data field is only allowed to be 0 or 1), the Pearson correlation is a universally understood, easy-to-compute statistic.

4.6.1.2 Ordinal Correlations

Ordinal correlations are an alternative to Pearson correlation that is more robust to outliers and more general in their underlying assumptions. Pearson measures the degree there is a *linear* relationship between X and Y; your data lie on a straight line. Ordinal correlations only measure whether the relationship is *monotonic*, with Y going up (or down) consistently for higher X. The plot in Figure 4.7 will have a Pearson correlation less than 1, but ordinal correlation of exactly 1.

Essentially they measure the degree to which sorting your datapoints by the X value is the same as sorting them by the Y value. They are necessarily robust to outliers because they rely only on the sorted order of the datapoints, not the actual values of X and Y. It doesn't matter whether the largest X value is 1% greater than the runner up or 100 times greater.

There are two types or ordinal correlation that you will see in practice, known as Spearman correlation and Kendall correlation. I have not personally had a time when the Kendall and Spearman correlations gave substantially different results, but at least in theory Kendall correlation is more robust to the occasional gross outlier (such as the tallest person in your dataset also weighing the *least*), whereas Spearman correlations are more robust to more modest variations.

The Spearman correlation is about the simplest ordinal correlation you could come up with. You simply replace every x value with its percentile rank among the Xs, and similarly with the Y values. Then you compute the Pearson correlation between the percentiles. The Kendall correlation is more complicated, and I won't get into the details here. The Spearman correlation takes considerably less time to compute, and for that reason if nothing else I prefer it in my own work.

4.6.2   Mutual Information

"Mutual information" (MI) is similar to correlation in that it quantifies the strength of the relationship between two variables, but it can capture relationships that are not monotonic. In theory the MI between X and Y tells you how much knowing the value of X allows you to make predictions about Y, relative to not knowing X. In practice though computing the MI requires a lot of strong assumptions about what form the relationship between X and Y can take, and often leads to thorny issues of computational tractability. So MI is often used as a way to identify features that are worth exploring further, but it is rarely used as a summary statistic for a dataset.

The biggest practical advantage of MI is that it applies to categorical data as well as numerical data, whereas correlation only applies to numerical data (or at least, in the case of ordinal correlations, data that can be sorted from "smallest" to "largest" even if it's not strictly numerical).

4.7   Advanced Material: Fitting a Line or Curve

In descriptive analytics settings the correlation between X and Y is often a good way to summarize their relationship; you want to know how good an indicator is of the other and your machine learning model will take care of the rest. You don't care particularly what the relationship between X and Y is, just whether it is strong enough to be useful or worth further investigation.

Other times though we want to go beyond gauging the strength of a relationship and nail down what that relationship actually is. Specifically, we want a mathematical formula that takes in X and yields a prediction for Y. This is critical if we want to do any sort of forecasting. Finding these formulas is a delicate balance between matching the available and thinking about what the relationship should look like.

In the simples case we might assume the linear relationship between X and Y; in this case we want the slope and intercept of the line of best fit. In other cases you don't expect a linear relationship in the data, and you will need to do something besides fitting a line for modeling. Some good examples include the following:

- One of your variables is time and you expect some kind of long-term behavior. An exponential decay to zero, hitting a stable plateau, etc. In these cases you are more likely to ask how quickly it converges to the final state and what that state looks like.
- There should be a point where the relationship is inverted. The more medicine you give a patient, the better they will be – until the side effects of the medicine become worse than what they're treating and the patient does worse. In this case you may want to estimate the "sweet spot" and how much wiggle room there is around it.
- You expect a feedback loop of exponential growth, like a website going viral. In this case you will ask how fast the exponential growth is (doubling every day? Every week?) and what level it starts at.

All of these fall under the umbrella of "parametric curve fitting." You assume there is a particular mathematical formula that relates x and y – a line, an exponential decay, etc. That formula has a small number of parameters that characterize it (the slope and intercept of a line, the growth rate and initial value of an exponential growth curve, etc.). You then try to find the values of those parameters that best match the data – those fitted parameters are what you use to summarize the data.

For simplicity imagine that we are fitting a line – the rest of this section applies equally well to an exponential decay curve, exponential growth, a parabola, or whatever else you may fancy.

Let me dip into some mathematical terminology. We say that the ith "residual" is $r_i = y_i - (mx_i + b)$; it is the difference (which can be positive or negative) between the height of the ith datapoint and the height of the line-of-best-fit. You can visualize the residuals in Figure 4.8.

There are two big questions to ask when fitting a curve: (i) how good is the fit, and (ii) was a line actually the right curve to fit? Residuals play a key role in answering both of those questions.

The typical way to measure how far off we are from the line of best fit is by assigning a "cost" to each residual, that is 0 if the residual is zero and positive otherwise, and then adding up the cost of every residual. Traditionally the cost of a residual is just its square, so that

$$\text{Total cost} = r_1^2 + r_2^2 + \ldots + r_n^2$$

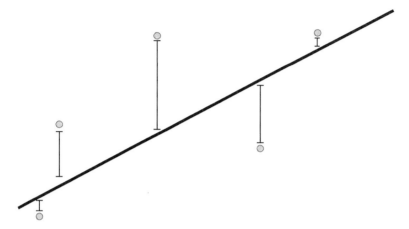

Figure 4.8 Residuals measure the accuracy of a model. Here the gray points are our data and the line is a line-of-best-fit through them. The residuals are how far the *y*-values in the data are off from the *y*-values predicted by the line.

Typically we divide this by what the total cost would have been if we had just passed a flat line through the average height of the data, as in Figure 4.9. Passing a flat line like this is about the most naïve "curve fitting" possible, so it is used as the benchmark for the penalty that a very bad fit would incur – getting less error than this means we are doing something right. This ratio – the total cost to what the cost would have been with a flat line – is called the "R squared" for the curve. It ranges from 0 to 1, with 0 meaning a perfect fit and 1 meaning that it did no better than the flat line.

Residuals can also be used to assess whether the model we fitted was even a good one to use, with a number called the "correlation of residuals." If the sizes and signs of the residuals are all mixed up, then they probably come from genuine noise in the data. But if there are large stretches where the residuals are positive, and others where they are negative, then this may be because we are fitting the wrong kind of curve. The distinction is shown in Figure 4.10.

If the residual for a given point is highly correlated (Pearson, Spearman or Kendall) with that of its neighbors, then this is a sign that we may want to revisit our assumptions.

Finally, residuals are the key to understanding how the best-fit parameters are actually found. Look back at the cost function earlier

$$\text{Total cost} = r_1^2 + r_2^2 + \ldots + r_n^2$$

where we add up the squares of the residuals to see how far off our line is from the data. If our dataset is set in stone, then this total cost is determined by the slope and intercept of our fit line, and we can find the "best fit" by finding the slope and

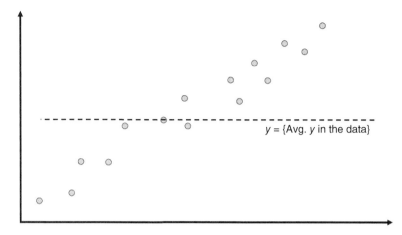

Figure 4.9 A degenerative form of "curve fitting" is used as a base of comparison for measuring how good a fitted curve is. In this baseline the value of the curve is always the average of all observed *y* values.

intercept that minimize the total cost. Because the cost of a residual is its square, this is called "least squares fitting," and it is by far the most common way to fit lines and other curves.

4.7.1 Effects of Outliers

I have mentioned several times that fitting a line is sensitive to the effects of outliers. This is why: the cost of a residual is its square, and that amplifies the effects of outliers. A single residual of size 10 is as costly as 25 residuals of size 2, so least-squares fitting will make a trendline in the name of accommodating a small number of outliers.

So what are we to do in situations where outliers are to be expected? One popular alternative to least-squares fitting is to say that the cost of a residual is simply its absolute value – outliers are still costly, but we don't amplify their impact by taking a square:

$$\text{Total cost} = |r_1| + |r_2| + \ldots + |r_n|$$

Why is it that least-square is so ubiquitous? The answer to this is mostly historical. For most cost functions you might select (including the absolute value) and you must find the best-fit parameters by a process called "numerical optimization": the computer takes a guess at the parameters, computes the total cost associated with them, and then gradually adjusts its guesses to get the cost lower. Eventually the guesses converge on a single solution, and there are some criteria for deciding that you are close enough. This is a very powerful process, but it

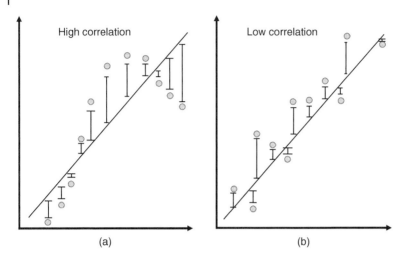

Figure 4.10 Large residuals can come from two sources: either that data we are trying to fit a curve to is noisy or we are fitting a type of curve that is a bad match for data. If nearby residuals are highly correlated with each other, it means that we have large spans where we are under- or over-estimating the function, suggesting that we've chosen a bad curve to fit. Uncorrelated residuals suggest that the data is simply noisy.

absolutely requires having a computer. In the special case of least-squares though, it happens that there is also a formula you can just plug your data into to get the exact best-fit parameters. Hence least-square was the only game in town in the days before computers, and the way to deal with outliers was to manually exclude them.

Now that we have computers it is possible to use domain expertise to specially tailor cost functions to the situation at hand. But as with all tools that offer extreme flexibility, it is possible to waste time perfecting your cost function while losing sight of what you're hoping to do with your fitted line. Unless there is a compelling reason to use something else, you should at least start with least squares – it is the easiest to compute and to communicate with people about. As always I say: use the simplest analytical technique that adequately solves the problem at hand.

4.7.2 Optimization and Choosing Cost Functions

As an aside I would like to say a bit more about numerical optimization, where we craft a function for the "total cost" and the computer finds the parameters that minimize that cost. We have talked about it in the context of fitting a curve, but the concept can be applied to virtually anything where you need a numerical answer. Optimization is an extraordinarily powerful technique, and it creates a natural dividing line between areas of specialty: you need domain expertise and business understanding to craft the cost function, but then you don't have to worry about how the optimal parameters are found (which software is used, whether it's on one machine or a cluster, etc.). Similarly a data scientist or engineer doesn't need any domain expertise to minimize a particular cost function once it has been crafted; for them it is just a well-defined math problem and they can find the best parameters in whatever way is most expedient.

Optimization is ubiquitous in numerical computing, and it plays a starring role in machine learning and artificial intelligence. When we talk about "training a model to data," what's really going on under the hood is numerical optimization. I will get into this a little bit more in future chapters, but for now I would like to give you a brief taste of things you can do with a carefully crafted total cost function:

- We discussed having the cost associated with a residual just being its absolute value, but then we still get punished for having large outliers (just not as badly as if we had taken the square). Instead you can pick a cost function that tops out at some maximum value, effectively giving a fixed penalty to large outliers regardless of how far out they are.
- Let's say that some datapoints are more important than others. We can multiply the cost of each residual with some importance weight before adding them up to the total cost.
- Let's say you don't care how big the residuals are in absolute terms, but how big they are relative to the correct answer. This would arise, for example, when you're trying to predict the movement of a price of stock. Then you could use a cost function like $(f(x) - y)/y$.

4.8   Statistics: How to Not Fool Yourself

So far in this chapter we have focused on ways to find, visualize, and quantify patterns in data. The problem though, which we have so far not discussed, is that sometimes those "patterns" are liable to be just flukes and there are severe limitations to how much we can conclude with confidence. Say, for example, that we have a coin that is biased, and we are trying to assess how biased it is. We flip it twice and get two heads. How much can we really conclude from this? If additional flips cost USD10 000 apiece, how many times are we willing to flip it?

There is a human tendency – especially in high-stakes situations – to grasp at a straw of signal in the data, concoct a story for why it must be legitimate, and move blindly forward with it. Statistics guards against this tendency; it is the science of quantifying just how confident we can really be. Sometimes decisions must still be made based on incomplete information, but at least you know what you know. And if you have the luxury of running controlled experiments, statistics tells you how much data is required to get the confidence you need.

Statistics is best understood in historical context. It was developed to deal with situations where data was sparse and costly, like the following:

- You want to know whether a new fertilizer helps crops. Every datapoint will cost you one plot of land and you'll have to wait a whole growing season to get the results back.
- The census has already been taken and there were 100 people. This is all the data you're getting.
- You are testing hardware prototypes in preparation for large-scale manufacturing. Every prototype will take hours to make by hand, and you'll have to pay for the parts.
- You want to know whether a new drug works. Every datapoint could cost you one human life.

In these situations you bend over backwards to eke all of the confidence you can out of every datapoint. Typically you don't have enough data to fit any kind of fancy predictive model; you're really just trying to tell whether a pattern in the data is a real-world phenomenon that you can make decisions on, or whether it's just a fluke because there's not much data. This assessment is made with something called a "*p*-value," which I will discuss in a minute.

Modern data science typically focuses on situations where this is not a problem. I often say that if you have a million datapoints and your pattern is weak enough that you even have to wonder about the *p*-value, then you can be certain that it's too weak to be of any practical use. The problem isn't sorting real-world patterns out from flukes in the data, but determining which patterns are strongest and most useful. For this reason data scientists are often ill-equipped to cope with the situations where statistics is still necessary. All of the areas that statistics was developed for still require it; human lives and plots of farmland are as costly as they ever were. It is the new data sources, like the Internet, where datapoints are cheap.

Statistics is a subject where seemingly simple problems can end up requiring surprisingly complicated math to deal with in a rigorous way. For that reason a significant amount of this section is labelled as "advanced material," which gets into the weeds further than most executives really need. It is here as a reference and help for building critical analytics thinking, but you're likely to be able to get by with just understanding *p*-values.

4.8.1 The Central Concept: The *p*-Value

Say that a disease has 50% mortality, and of the 10 people to whom you gave a new drug 8 survived. Can you be confident in this case that the drug works, or did the extra three patients just get lucky?

For illustration purposes I'd like to discuss something a little simpler: you've flipped a coin 10 times and gotten 8 heads, and are wondering whether the coin is biased. The first thing to understand is that we are not asking for "the probability that this is a biased coin." Gauging the actual risk of a biased coin involves a subjective assessment that is extraneous to the data – those 8 heads will look much more suspicious at a seedy casino than with a coin from your own pocket. We are looking for something a little more objective.

Making statements about what the world is probably like requires subjective assessments. Instead statisticians go the other way: they make mathematical assumptions about how the world is, and then calculate how likely certain things about the data are *given* those assumptions. In this case we take for granted that the coin is fair and we ask: what is the probability of getting 8 or more heads out of 10 flips? In this case it is 5.4% – small, but not miniscule. You can then weigh this objective metric with any subjective judgment calls you might want to make.

In statistics terminology the notion of a fair coin is called the "null hypothesis." The null hypothesis means that there is no pattern in the real world – the coin is fair, the medicine doesn't affect patients, etc. We don't necessarily believe the null hypothesis is true (we usually hope it's not!), but it gives the mathematical precision that is required for computing hard numbers. According to the null hypothesis, every pattern we see in the data is just a fluke.

The fraction of coins that came up heads is called the "test statistic" – it is a number that we compute from our data that can be used to gauge the accuracy of the null hypothesis. The null hypothesis's mathematical precision allows us to do one key thing: compute *exactly* how likely the test statistic is to fall within a given range. If the test statistic falls well outside of the expected range, then the null hypothesis starts to look pretty dubious.

In the case of coin flips, it is natural that we would use a fair coin as the null hypothesis, and our test statistic would be the fraction of flips that came up heads. There's really no other way to do it, so you might wonder why I'm introducing this additional terminology. We will see though that in other cases it's not so obvious what the null hypothesis should be, and picking the right test statistic is not so easy.

The "*p*-value" – the central concept of statistics – is the probability of getting a test statistic that is at least as extreme as the one we observe *if* you assume the null hypothesis is true.

It is a convention to take 5% as the cutoff for whether a pattern is "statistically significant," but that's entirely convention. And it's worth noting the converse; if the null hypothesis is true and there really is nothing to see here, then 5% of the time you will still get a "statistically significant" result by random chance.

The key asterisk to keep in mind is this a low *p*-value does not mean a strong pattern. The process I've outlined here (which is called "hypothesis testing") tells you how plausible the null hypothesis is given your data, but it does not tell you how badly the null hypothesis gets violated. It is often the case that the underlying pattern is quite weak, but you just have so much data that it is undeniable. Take the null hypothesis of a fair coin, for example. The *p*-value for getting 8+ heads from 10 flips is 5.4%. But if you flip 100 times and get 60 heads (a weaker signal, but with more data), the *p*-value goes down to 2.8%. Flipping a slightly biased coin many times can give the same *p*-value as fewer flips of a more egregiously biased one. As datasets get larger more and more tiny, effects become statistically significant, and you switch from asking whether a pattern is real to whether it is strong enough to be useful.

4.8.2   Reality Check: Picking a Null Hypothesis and Modeling Assumptions

A key limitation of *p*-values is that they depend on the mathematical structure of the null hypothesis, and often that requires very strong assumptions. The coin flip example is a little artificial because the assumption of a fair coin is quite simple mathematically and it's really the only natural choice. Often in order to compute a *p*-value, you must make much stronger assumptions than "there is no pattern," frequently about data following a particular distribution (usually the normal distribution). These assumptions are made up out of whole cloth; they are necessary in order to do the math, but they introduce a frustrating uncertainty into the process.

For example, let's say that the salespeople in office A make more sales on average than in office B, and you want to know whether that's just coincidence (one office would have a slightly higher average just by dumb luck) or whether A is genuinely a better office.

The revenues from *A* and *B* both follow probability distributions, and the null hypothesis is that the average of those distributions is the same; the observed difference between their revenues was just a statistical fluctuation. Fair enough, but that's not enough to compute a *p*-value. Intuitively, in order to know how large a fluctuation is typical, we must also know how much variability there is in *A* and *B*'s distributions. And what if the averages are the same, but one office has more variability than the other? Ultimately to run the math we have to assume what type of probability distribution *A* and *B*'s revenues follow; the near-universal convention is to assume they are normal distributions.

You have almost certainly heard warnings about how poor a model normal distributions are. In this case the biggest problem is that they have very few outliers, because it is very possible for an office to have abnormally good or bad years. So should we instead go far off the beaten path and use a different probability distribution? There is no bottom to this rabbit hole; ultimately you just have to pick something and hope that it's accurate enough.

Instead of embracing fancy math and dubious modeling assumptions, I usually recommend trying to structure your analysis in a way that minimizes those complications in the first place. Measuring an office's sales is tricky because you have to make modeling assumptions about probability distributions that are probably false. Things become much more tractable if you use a binary metric, like whether or not a person's sales exceeded a certain performance threshold. In that case having 100 people in the office is equivalent to flipping 100 coins, and your null hypothesis is that each office's coin has the same chance of coming up heads. Of course measuring the percentage of employees who are "good enough" is different from measuring the overall average sales, but for many business decisions these metrics might be equally defensible.

In data science you will often see this in the context of A/B testing. Usually something is either a success or a failure determined by whether or not a user engages in some target action (like making a purchase on a website). Repeatedly flipping a biased coin is an extremely accurate way to model these situations, meaning that our p-values will be very reliable. When we get into situations with normal distributions, I still compute p-values, but I also start to get nervous.

Because of these limitations it is important, when analyzing experimental data, to still present the data graphically in addition to giving the p-value. Showing histograms of the sales from offices A and B will often make the statistical conclusions visually obvious.

4.8.3 Advanced Material: Parameter Estimation and Confidence Intervals

You've seen people reporting that the average height of a population is, say, 170 ± 5 cm. This means that a sample of the overall population was taken; the average height of people in the sample was 170 cm, and the average height over the entire population is probably between 165 and 175. It could be outside that range, but it's considered unlikely.

The range from 165 to 175 centimeters is called the "confidence interval." There are two key things to know about confidence intervals around the average for data that is approximately bell-shaped:

1. The size of the confidence interval is proportional to the standard deviation of the sample. If there is twice as much variability between individuals, there will be twice as much uncertainty in the what you can estimate from a sample of those individuals

2. The size of the confidence interval goes down as the square root of your number of datapoints. If you get four times as many datapoints, the interval will shrink by a factor of 2. A hundred times as much data will shrink it by a factor of 10 and

so on. As you get more and more data, you can make the confidence interval arbitrarily small.

A vast majority of the time, this will stand you in good stead. In my experience people are pretty casual in how they use confidence intervals – often they will literally just divide the standard deviation of the heights in the sample by the square root of n and call it good.

Strictly speaking though there is a bit more to computing confidence intervals. Let's back up for a second and frame things more mathematically. In the example of heights, we are tacitly assuming that the heights of people in the overall population are described by a normal distribution, whose defining parameters are its mean μ and standard deviation σ. The values of those parameters are unknown, and our goal is to estimate them from the heights in our sample of n people.

The confidence interval is the set of parameter values for which, if you take that value as a given, the observed data is considered plausible. When I said earlier that "the average height over the entire population is probably between 165 and 175," I was technically lying. This isn't an issue of probabilities: the average height is either within the confidence interval or not. The confidence interval from 165 to 175 instead represents the collection of values of μ for which the observed data would have a p-value greater than 0.05.

Computing the confidence intervals for a particular parameter of a particular probability distribution is in general quite complicated. It just happens that for normal distributions the confidence interval is approximately equal to the standard deviation in your sample times the square root of n.

This distinction can become relevant when you're estimating parameters for something other than a bell-shaped curve. For example, let's say that you are flipping a biased coin and trying to estimate the probability p that it comes up heads. We will say that getting heads is 1 and tails is 0. You flip it four times and get four heads. If we naively carried over lessons from the normal distribution we would conclude that $p = 1.0$ is our best estimate of the parameter, and our confidence interval has zero width since the standard deviation in our samples is the same. But clearly a biased coin with $p = 0.9$ would still be entirely plausible given this data! The correct, rigorously derived confidence interval (which is called the Clopper–Pearson interval in this case) takes this into account and is approximately

$$0.3976 \leq p \leq 1.0$$

4.8.4 Advanced Material: Statistical Tests Worth Knowing

This section will give a brief overview of several specific statistical tests that are worth knowing. For each one I will explain intuitively the null hypothesis that you are trying to test, discuss the test statistic that you compute to test it, and give some general pointers.

4.8.4.1 χ^2-Test

The χ^2-test (pronounced "chi-squared") is used when we are measuring a categorical variable, like flipping a biased coin, rolling a dice, or getting a state in the union. Each possible outcome has a fixed probability (the outcomes are equally likely in the case of rolling a dice, but in general they can be different), and we want to test whether those probabilities are indeed the correct ones. Essentially this is a goodness-of-fit test between the frequencies that we believe are true and the data we observe.

We expect the values that we think are common to occur a lot in our data, and the values that we think are rare to occur only sporadically. Specifically if the ith outcome has probability p_i and there are n datapoints, then you expect the number of times it occurs in the data (which we will call d_n) to be approximately $n*p_i$. The test statistic you compute combines the deviations from all of these expectations:

$$\sum_i \frac{|x_i - n*p_i|^2}{n*p_i}$$

The χ^2-test gets its name because – if the null hypothesis is indeed correct – then the probability distribution of the test statistics is approximately what's called an χ^2-distribution.

Note one limitation of the χ^2-test. Say we are rolling a weighted dice many times, and we expect that it only comes up as a five 0.01% of the time – almost never. If the actual probability is 1% then x_5 will be many times larger than np_5, the test statistic will be very large and the test will fail. And of course it *should* fail because the null hypothesis isn't technically correct, but for some business applications we only want it to be "correct enough" and those highly-unlikely faces are kind of inconsequential. In such situations you can group together the least likely outcomes into a single category and re-run your test.

4.8.4.2 *T-test*

I previously mentioned the situation of comparing the average sales revenue between offices A and B in order to determine whether the higher-earing office is truly better or just got lucky. In that case we model the revenue from a given employee as a bell curve with mean μ standard deviation σ, and we are wondering whether μ is the same for both offices. This is not too hard to compute if you know what σ is and you can assume that it is the same for each office. But generally σ itself is unknown (and is generally different between the two offices) and will have to be estimated from the data.

What is called Student's T-test is used to cope with this situation. Essentially you fit a normal distribution separately to the data from each office, computing best-fit estimates for its μ and σ. You then look at how large the difference between the μs is, and scale that difference down using the σs. This spread-adjusted difference between the average sales is your test statistic, and if the null-hypothesis is true, it is follows a probability distribution called Student's T-distribution.

The biggest issue with this T-test is that it tacitly assumes that the sales for an employee in a given office follow a normal distribution and assumptions like that are typically not true. If you assumed a more appropriate underlying distribution, it would be possible to derive a different test statistic and p-value, but you will almost never see this. Instead people generally take T-tests with a pinch of salt, and make sure not to apply them to data that is too heavy-tailed.

4.8.4.3 Fisher's Exact Test

Fisher's exact test is used to determine whether the proportion of one group who meets some criteria is greater than another group, especially in situations where there is extremely little data available. For example, say that we have a group of men and women, each of whom either liked or disliked a movie, as shown in this table:

	Men	Women	Row total
Liked the movie	1	9	10
Disliked the movie	11	3	14
Column Total	12	12	24

In this case $1/12 = 8.33\%$ of men liked the movie and $9/12 = 75\%$ of women liked it. This is a stark difference in percentages, but the numbers are so small it is not immediately clear whether it could be a coincidence.

The null hypothesis of course is that neither gender has more of an affinity for the move than the other. In Fisher's test we say that there are 24 datapoints each of which is given a gender and a movie status. There are only so many ways they can be apportioned so that there are 12 women and 10 people who like the movie. The test statistic is this: in what fraction of them do at least 75% of women like the movie? In this case the p-value is approximately 0.0013 – despite the paucity of datapoints we have an extremely significant result.

4.8.4.4 Multiple Hypothesis Testing

I have alluded several times to the risk that if you look hard enough in your data, you will eventually find something that is statistically significant, even if just by chance. Let's look at that a little more closely.

If the null hypothesis is true, then you have a 5% chance of getting a fluke in your test statistic and erroneously declaring a result to be statistically significant. If you run two tests for which the null hypothesis is true, and those tests are independent, then the chance that *at least* one of them is statistically significant is $1 - 0.95 \times 0.95 = 9.7\%$, higher than if you only tested a single hypothesis. If you run 14 tests all of which have no real pattern, then you will *probably* (~51% chance) have at least one that is statistically significant by chance.

This is the subject of "multiple hypothesis testing," and it comes up whenever there are multiple test groups in a controlled experiment. Intuitively the way to do this is to tighten our criteria for what counts as statistically significant; if you are testing ten hypotheses rather than just one, you set a higher bar than you normally would before you declare any finding to be statistically significant.

Dealing with this rigorously is quite complex if there is any correlation between the various tests you are running. Generally speaking people assume that each test is independent and apply what is called the "Bonferroni correction." You divide your confidence threshold (typically $p = 0.05$) by the number of different tests that you are running to get a much tighter threshold $p^* = \frac{p}{\{\#tests\}}$. You only declare a result to be statistically significant if its p-value is less than p^*, rather than p. You can show mathematically that if all of your null hypotheses are true, then the probability of *any* erroneous discovery is then the original p.

Besides analyzing data after-the-fact, this correction becomes very important when designing experiments. Typically when you design an experiment, you decide how large of an effect you want to be able to detect, and then gather enough data that such an effect would probably yield a p-value below your confidence threshold. But say that you have five different success criteria that you are looking at, and you want to see if *any* of them becomes more common in the test group than the control group. The Bonferroni correction means that you will use a p-value of 0.01 rather than 0.05 when it comes time to analyze the data. Assuming there is a pattern, as you gather data your p-value will reach 0.05 before it gets down to 0.01, so you will need to gather more data before you are able to declare that any single success metric has gone up significantly.

4.8.5 Bayesian Statistics

As I discussed earlier, p-values don't tell us how likely the world is to be a certain way – they tell us how likely our data is under given assumptions. Making statements about the world requires a lot of messy judgment calls and subjective assessments. The point of statistics is to remove human subjectivity so that we can put our preconceived notions to the acid test. This is called "classical" or "frequentist" statistics.

There is another school of thought called Bayesian statistics that (for better or worse) embraces preconceived notions. It is a way to leverage high-quality human expertise in situations where the data is extremely sparse. It is inspired by the notion that we start off having a set of hypotheses about the world, with varying degrees of confidence in them, and we update our confidence in the different hypotheses as we gather more data.

In classical statistics we typically have a single null hypothesis, and we are assessing whether the data is plausible given that hypothesis. In Bayesian statistics we have several hypotheses about the world, exactly one of which is true, and some level of confidence in each of them. These confidences are called "priors." The difference between "confidence" and "probability" is a matter of philosophical debate – I will just say that they are mathematically equivalent and move on.

As data becomes available we update the odds of one hypothesis versus another by how likely the data is under those hypotheses. You can show mathematically that the correct formula to update the odds is the following:

$$\frac{\text{Prob}\{A\}}{\text{Prob}\{B\}} \leftarrow \frac{\text{Prob}\{A\}}{\text{Prob}\{B\}} * \frac{\text{Prob}\{\text{Data} \mid A\}}{\text{Prob}\{\text{Data} \mid B\}}$$

Rather than a few discrete hypotheses about the world, Bayesian statistics also lets us have a continuum of hypotheses with varying levels of confidence. For example, you can have a bell-shaped curve that represents your prior confidence about where the "true" average of data in the world is likely to be. As data comes in your bell curve shifts around and get more pointed: the ultimate bell-curve will be a balance between your prior confidences with the data that has come in, tilting toward the latter as more data comes in.

Bayesian statistics is more mathematically complicated than the classical approach. Both of them require computing the probability of observed data given an underlying hypothesis about the world. But Bayesian statistics also requires prior confidences, which can be a can of worms in themselves, and your final probabilities will be a mish-mash of the two. This can be worth the effort if you have high-quality priors and not a lot of data. If there is a lot of data available though, classical and Bayesian statistics quickly start giving answers that are functionally equivalent.

4.9 Advanced Material: Probability Distributions Worth Knowing

I've spent a lot of time pointing out the shortcomings of the bell-shaped curve and analytics that is based on it. The bell curve is an idealization resting on very strong mathematical assumptions that often don't hold up in the real world. But there are other probability distributions out there that are based on different assumptions, which may be more applicable to a given situation. No model is perfect but some of them are quite accurate, and modeling is a requirement if we are to make probabilistic predictions, calculate odds, or apply statistics. There is much more to the world of probabilistic models than the classical bell curve: the goal of this section is to give you an overview of it.

The key thing to understand about probability distributions is that most of the interesting ones will arise naturally in certain idealized situations. For example, if you add up many *independent* random numbers (flips of a coin, transactions that generate some amount of revenue, etc.), their sum is necessarily well-described by a bell curve; if the numbers are *not* independent then the bell curve breaks down. Every other probability distribution in this chapter has a similar story of how it could arise, and knowing how the different distributions arise lets you guess which ones will be promising (or dangerous!) for describing a particular situation. Don't think of the probability distribution as a mathematical object; it is a narrative describing the real-world process that generated the data.

This section will be somewhat mathematical; obviously a probability distribution must be treated as math when it comes time to compute a number. But I will try to keep those details to a minimum. Instead I will focus on how the different distributions arise, typical real-world situations they are used in, and the key assumptions that they hinge on. This is what you need in order to thinking critically about modeling decisions.

4.9.1 Probability Distributions: Discrete and Continuous

A probability distribution is a general, mathematical way to say which numbers are how likely to happen. They come in two main varieties:

- *Discrete distributions*: The data can only come in discrete integers, and each one has a finite probability. Rolling a dice is a good example, where each of the numbers 1–6 has probability one sixth. Another is flipping a biased coin, where perhaps the probability of getting a head (which is usually described with the integer 1) is 2/3 and a tail (described with 0) is 1/3. Coins and dice rolls both have a finite set of possible outcomes, but that is not a requirement. You would also use a discrete distribution to model the number of visits a website receives in a year; in theory it could be arbitrarily large, but the likelihood of the large numbers becomes vanishingly small.
- *Continuous distributions*: In this case the data are decimal numbers, not integers. Any single number has probability zero, but we can say how likely a datapoint is to lie between any two numbers.

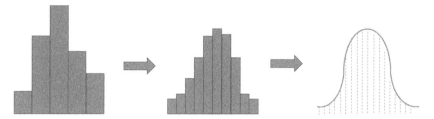

Figure 4.11 The most intuitive way to think of a probability distribution is as a histogram with very narrow bins and a very large amount of data in it.

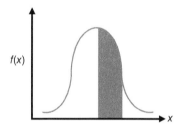

Figure 4.12 The area under the curve of a continuous probability is distribution is the probability that a value will be somewhere in that range.

The relationship between discrete and continuous distributions is somewhat fluid. For example:

- The number of M&Ms in a bucket is technically described by a discrete distribution, but for most practical purposes it probably makes more sense to describe and measure it in terms of weight, which is continuous.
- The number of visitors to a website in a day is a discrete distribution. But for analysis purposes we will often take a moving average to remove noise, at which point "2.278 people" becomes a meaningful concept.
- Computers can only store numbers up to so many decimal places, so in this sense computers only work with discrete distributions.

Most continuous distributions can be thought of as the limiting cases of some underlying discrete distribution, like M&Ms in a bucket. I will highlight when this is the case.

The best way to think of a probability distribution is as a histogram with a massive number of datapoints in it (and extremely thin bars in the case of a continuous distribution), so that all the noise smooths out. The heights of all the bars are scaled so that their areas add up to 1. You can visualize it as in Figure 4.11.

The probability of a particular value in the data, or range of values, can then be found by looking at the area under the probability curve, as shown in Figure 4.12.

The focus of this section will be how these probability distributions arise and the properties they have. I will generally motivate this visually by examining what the probability distributions look like, rather than mathematically with equations.

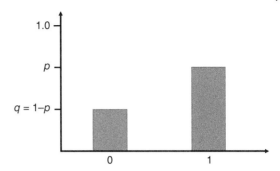

Figure 4.13 The Bernoulli distribution is just the flipping of a biased coin. It is 1 with probability p, and 0 with probability $1 − p$.

However I will note that for a discrete distribution we typically denote the probability of a number N by p_N, and for a continuous distribution we often say that the height of the distribution at x is $f(x)$.

4.9.2 Flipping Coins: Bernoulli Distribution

The simplest probability distribution is the so-called Bernoulli distribution, shown in Figure 4.13. It is simply the flipping of a biased coin, which has probability p of coming up heads. The probability of tails, $1 − p$, is conventionally called q. It is typical to identify heads with the integer 1 and tails with the integer 0.

The Bernoulli distribution can be used to describe any yes/no or other binary situation. Visitors to a website either ultimately make a purchase or they don't. Medical tests for antibodies to a virus either find antibodies or fail to.

The Bernoulli distribution is not particularly interesting in itself, but it is a building block for more complicated distributions like the binomial.

4.9.3 Adding Coin Flips: Binomial Distribution

A binomial is the number of heads when you flip n biased coins – the sum of n independent Bernoulli distributions. It is extremely common in data science as a way to model a collection of things that could either succeed or fail.

In particular it is very useful in A/B testing, where we randomly assign users to the text or control group and look at the proportion that succeeds. The number of successes in the test group will be binomially distributed, except that we don't know the value p of success.

The most important thing about the binomial distribution is that as n gets bigger, the so-called law of large numbers kicks in – the number is almost certainly close to its average of $n*p$. If you flip a hundred fair coins, then there will probably be close to 50 heads. If you flip a thousand coins then it will almost certainly be close to 500 heads, and so on.

To be a little more precise about this, the standard deviation of the binomial distribution is $\sqrt{p(1-p)} * \sqrt{n}$. This means that as n grow the standard deviation is proportional to the square root of n. But the average pn is proportional to n itself, so the standard deviation gets smaller and smaller relative to the average. To illustrate this numerically let's flip a fair coin with $p = 0.5$, so that $\sqrt{p(1-p)} = 0.5$. Then if $n = 100$ the standard deviation will be $0.5 \times 10 = 5$, i.e. which is 5% of the average. But if $n = 10\,000$ then the standard deviation will be $0.5 \times 100 = 50$, which is just 0.05% of the average. As n grows larger the probability of a meaningful deviation from the average becomes negligible.

The key assumption behind the binomial distribution is that all of the coins are independent of each other. In particular there is no outside force that biases large swaths of them together toward heads or tails. This assumption is crucial for the dearth of outliers in the binomial distribution.

The financial crash of 2008 is an excellent example of this independence assumption breaking down. I will simplify things a bit for the sake of illustration. Say that you have a portfolio of one thousand high-risk USD100 000 mortgages, each of which has a 75% chance of getting paid off (in which case you get the full USD100 000) and 25% chance of defaulting (in which case you get). So you can get a maximum of USD100 million if everybody pays their mortgage, but the average is USD75 million. The number of people who pay their mortgages can be modelled as a binomial distribution, with $n = 1000$ and $p = 0.75$.

Say further that you want to turn these high-risk mortgages into low-risk financial instruments that you can sell. The way to do that is to sell two bonds: bond A that receives the first USD50 million in mortgage payments and bond B that receives whatever else comes in (maxing out at USD50 million, but typically running around USD25 million). In financial parlance these are different "tranches" of bonds. The risk of bond A being worth anything less than the full 50 million is the probability of this binomial distribution being less than 500 – less than 1 in a billion. If our probability estimates are off and the mortgages only have a 70% chance of getting paid, it will be bond B that suffers. Bond B essentially acts as cannon fodder that allows bond A to be low risk, even though the underlying mortgages have a 25% chance of defaulting.

However, this all hinges on the underlying assumption that the mortgages are independent. If those thousand mortgages all come from the same company town, and that company starts having layoffs, then the assumption of independence goes out the window. Bond A is no longer protected by the law of large numbers; it will falter if the company falls on hard times, regardless of how big n is. Note that this can still happen even if 75% of similar mortgages in the country get paid in full; correlation between mortgage payments is enough to turn make bond A risky, because the variability in mortgage payouts increases.

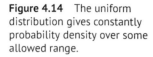

Figure 4.14 The uniform distribution gives constantly probability density over some allowed range.

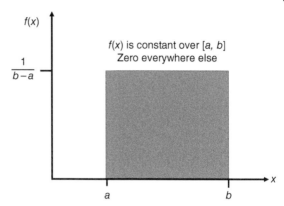

You can have safer bets with something like life insurance. Recessions are a known phenomenon that makes risks become highly correlated. On the other hand massive die-offs are much less plausible in modern times (although cross your fingers regarding disease epidemics...).

4.9.4 Throwing Darts: Uniform Distribution

The uniform distribution is the simplest continuous distribution, shown in Figure 4.14. It is characterized by two numbers a and b; the number is equally likely to be anywhere between a and b, but cannot be outside that range. The uniform distribution is useful if we have some principled reason for knowing that a number must be within a particular range, but beyond that we know nothing.

4.9.5 Bell-Shaped Curves: Normal Distribution

The classical bell-shaped curve is also called a "normal distribution" or a "Gaussian," visualized in Figure 4.15. It is defined by its mean (the location of the top of the bell, typically denoted by μ) and standard deviation (typically denoted by σ, it measures the width of the bell). The normal distribution has extremely thin tails – large outliers are virtually unheard of. For this reason it is an imperfect way to model many real-world phenomena, but often good enough for the task at hand.

The normal distribution is ubiquitous in classical statistics for a number of convenient mathematical properties it has. By far the most important of them is the "central limit theorem":

Central limit theorem. Say you have a probability distribution with mean μ and standard deviation σ. If you take n independent samples from it and take their average, that average is approximately a normal distribution with mean μ and standard deviation $\sigma/\sqrt{n}$.

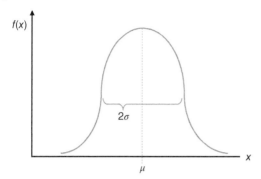

Figure 4.15 The normal distribution, aka Gaussian, is the prototypical "bell curve" and the most important distribution from classical statistics. It is characterized by its mean (the value where it peaks) and standard deviation (which measures how far values typically are from the mean).

In practice the average converges rather quickly most of the time to a normal distribution. Outliers and heavy tails in the distribution make no difference to the theorem – the convergence just happens more slowly.

In any application where multiple more-or-less independent numbers get added up, you should start thinking about a normal distribution. This includes the aggregate revenue that a salesperson brings in during a year, the garbage produced by an office during a day, or the lines of code an engineering team produces in a month. As with the binomial distribution thought, the big thing you should watch out for worry about is if the numbers are correlated with each other. The central limit theorem works because the datapoints above and below the mean tend to cancel each other out, but that ceases to hold if they are correlated.

4.9.6 Heavy Tails 101: Log-Normal Distribution

The log-normal distribution is probably the most common heavy-tailed distribution you see in practice. When many independent numbers are added together, their sum can be described by a normal distribution; if they are instead *multiplied* together then the *product* becomes log-normal. You should think about log-normal distributions any time that something changes proportionally to its value, such as:

- Investment returns over a period of time, where each day your portfolio value gets multiplied by a number indicating how well for did that day.
- The size of a corporation, which tends to grow or decline in proportion to how big it currently is.

The log-normal distribution has many convenient properties that make it an excellent match for real-world situations like net worth:

- It can take on any positive value, but is never 0 or negative.
- There is a single relatively clear peak.
- While the tails are heavy, the average and standard deviation are both finite.

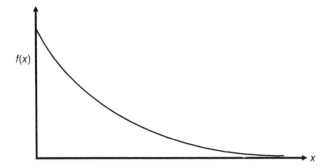

Figure 4.16 The exponential distribution is often used to estimate the length of time between consecutive events. Its most important property is that it is "memoryless"; if the first event happened T seconds ago and you are still waiting for the next event, the *additional* time you have left to wait still follows an exponential distribution. It is the continuous analog of the geometric distribution, which measures how long you have to wait until a biased coin comes up heads.

Mathematically you can sample from a log-normal distribution by sampling from a classical normal distribution with mean μ and standard deviation σ and then taking the exponent of this value. The value of μ tells you where the peak of the log-normal distribution is, and greater σ will create a heavier tail off to the right (note that the average value of a log-normal distribution will be to the right of the peak, because it average gets pulled up by this one-sided heavy tail).

The log-normal is my go-to distribution when I expect a situation to be heavy-tailed.

4.9.7 Waiting Around: Exponential Distribution and the Geometric Distribution

The exponential distribution is used to model the time between independent events, and its probability density is shown in Figure 4.16. The time between customers visiting a website, oil rigs breaking down in a field, and asteroids hitting the earth are all exponentially distributed. The exponential distribution is characterized by a single number: the average time between events.

The best way to think about it is this. Every minute in time that passes you flip a biased coin – if it comes up heads an event will happen, and in any case you then wait until the next minute and flip again. This is a discrete-time way to model whether events happen. Now instead flip a coin every second, but with a coin that is one sixtieth as likely to come up heads; this will give the same number of events in a day on average, but with finer granularity of time between events. In the limit of fine granularity the time between events becomes exponentially distributed.

My analogy of coin flips was deliberate: the exponential distribution is the continuous analog of the discrete "geometric distribution." The geometric distribution models how many times you have to flip a biased coin until it comes up heads. The exponential and geometric distributions have many of the same properties and applications – the difference between them in practice is often whether you measure time continuously or in distinct steps.

The key property shared by both distributions is that they are "memoryless." In flipping a coin it doesn't matter if the last head was 10 flips ago, a 100 flips ago, or a million: the number of flips you have left to wait until the next head is the same. Similarly in an exponential distribution it doesn't matter how long you've waited for an event to happen; the average time you have left to wait is the same as it ever was. If the time *between* the events is an exponential distribution, then the time *remaining* until the next event is the same exponential distribution.

Exponential distributions are often used to model the waiting time between events when studying a complex system. If you are trying to simulate the performance of a database that fulfills incoming requests, or the behavior of arriving at a supermarket checkout line, the exponential distribution is the go-to option. This runs into problems though because events often come in bursts – short periods of high intensity followed by lulls. Many event-driven simulations provide grossly inaccurate estimates because they fail to take the reality of bursts into account.

Another key use for the exponential distribution (which we will see more about in Section 4.9.8) is to model the time we must wait until a single inevitable event occurs. For example, we could model the lifetime of an office stapler using an exponential distribution. Every day there is a finite probability that somebody will lose it. The risk of losing the stapler on a given day does not grow or shrink as time goes on, and the age of the stapler has to bearing on how careful people will be with it today. There is no way to predict when it will occur, but eventually somebody will lose the stapler.

4.9.8 Time to Failure: Weibull Distribution

The Weibull distribution is a generalization of the exponential distribution, which is especially useful for modeling the lifespan of something (how long the office stapler stays around, how long somebody stays a loyal customer, etc.).

Go back to the example of the useful life of the stapler. The exponential distribution is great for modeling this – provided that the stapler's tenure ends when somebody loses it and hence there is no way to predict how much longer it will be around. But there are two other (contradictory) phenomena we might want to capture:

- *Infant mortality.* Some staplers come off the factory line defective and will break early before anybody has a chance to lose them. The longer a stapler has been working, the more likely it is to be one of the good ones. So in contrast to the memoryless exponential distribution, the longer a stapler has been around, the *more* time it probably has left.

- *Wear out.* The stapler wears out over time and becomes more likely to break with normal usage. So in contrast to the exponential distribution and to infant mortality, the longer a stapler has been around, the *less* time it probably has left.

So that is the key question we ask when modeling a situation: if the stapler has been around longer does that mean that on average it has more time left (infant mortality), less time left (wear out), or no difference (in which case it's an exponential distribution).

The Weibull distribution is like the exponential distribution except that there is an additional parameter that indicates how much infant mortality (or conversely, wear out) it has.

4.9.9 Counting Events: Poisson Distribution

The Poisson distribution, shown in Figure 4.17, is a discrete distribution that is used for modeling the number of events that happen in a period of time. It is characterized by a single parameter: the average number of events. There is a finite probability that zero events will happen, and technically arbitrarily many events

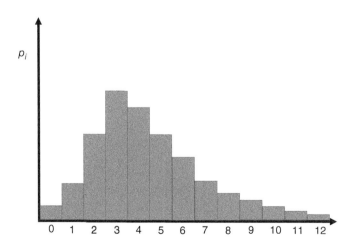

Figure 4.17 Say there are many independent events that *could* happen (there are 10 billion people in the world, all of whom *could* hypothetically visit your website in a given day), but each event is very unlikely and on average only a modest number actually *will* happen. The Poisson distribution is used to model such a situation. Its key assumption is that the various events that could happen are all independent of each other.

are possible, but in practice the probability mass is concentrated at a peak near the average and there are no heavy tails.

The best way to think about Poisson distributions is that there are many independent events that *could* happen, but they are all unlikely and on average only a small number of them will occur. The number of new telemarketers who cold-call you during a week is a Poisson distribution. So is the number of car crashes in the United States in a year.

The key assumption about a Poisson distribution is that the events are independent of each other. For example, you might want to model the number of visitors to a website on a day; there are several billion people in the world, but on average only a tiny fraction will visit your site. This works great if there is no coordination between people. But if one person can influence others, say by posting a link on Facebook, then independence breaks down and you will no longer have a Poisson distribution. Or if some external event is liable to influence large swaths of people, then your daily traffic will reflect whether or not that event happened.

Recall that a binomial distribution occurs if you flip n coins, each of which has probability p of coming up heads, and you count the resulting heads. On average there will be $p*n$ heads, but in theory there could be up to n. If you flip more and more coins, but you make p smaller and smaller so that the average $p*n$ stays fixed, then a binomial distribution becomes a Poisson distribution.

The Poisson distribution is closely related to the exponential distribution. If the time between consecutive events is exponentially distributed and those times are all independent of each other, then the total number of events the happen in a given time interval is Poisson-distributed.

Glossary

A/B **testing**	Randomly dividing things into a control a group and one or more test groups to see whether something we do differently in the test groups has any effect. Random partitioning is the key that allows us to distinguish correlation from causality.
Anscombe's quartet	A famous collection of four graphs that are visually obviously different, but that have identical trendlines and summary statistics.
Binomial distribution	A probability distribution you get by flipping a fixed number of biased coins and counting how many come up heads.
Bonferroni correction	If we are testing multiple hypotheses and want to see if any of them are statistically significant there is a greater risk that one will look significant by

chance. The Bonferroni correction adjusts our *p*-value threshold down to account for this risk.

Chi-squared test A statistics test that lets us determine whether the probabilities we have assigned to some discrete collection of outcomes (like flipping a biased coin or rolling a weighted dice) are indeed the correct probabilities.

Correlation Any of several metrics that measure the relationship between variables *X* and *Y* in data. The most popular choice is Pearson correlation, but ordinal correlations and more robust to outliers.

Correlation of residuals In curve fitting this measures whether the errors in the model come from noise in the data or from a poor choice of model. Highly correlated residuals indicate that the choice of model is a bad match for the data.

Descriptive analytics A collection of methods, both visual and numerical, to describe and summarize the contents of a dataset. It is in contrast to predictive analytics where we fit models in the hopes of predicting what future data will look like.

Exponential distribution A probability distribution often used to model the time between consecutive random events happening.

Gini coefficient A metric that measures how heavy-tailed a distribution is. It is especially used in the context of social science and studies of wealth distribution.

Heavy tails A situation when a probability distribution has a significant probability of outlier events.

Kendall correlation A popular ordinal correlation.

Log-normal distribution A simple probability distribution that displays heavy tails. It is often used to study situations where many *proportional* changes get aggregated together by multiplication, such as investment return.

Median The middle point for some value in a dataset, where half of data are above it and half are below. This measure is robust to outliers in a way that the average value is not.

Mode	The most common value observed in data.
Null hypothesis	In the context of statistics this is a mathematical version of the assumption that there is no real-world pattern.
Ordinal correlation	A correlation metric that measures the degree to which sorting data by X is the same as sorting it by Y. Ordinal correlations are robust to outliers because the actual values of X and Y are irrelevant. Spearman and Kendall are the most popular ordinal correlations.
Pearson correlation	The traditional correlation metric. It measures the degree to which the datapoints all lie on a fixed line and is sensitive to outliers in the data.
p-Value	The probability that a test statistic will be at least as extreme as the one we see in the data, according to the null hypothesis. A tiny p-value means that the data looks less plausible if we assume the null hypothesis, and casts doubt upon the null hypothesis.
Poisson distribution	A probability distribution used to model the number of independent events that occur in a time period.
Residuals	In curve fitting these are the difference between predicted values of Y and the observed values of Y.
Robust statistics	A collection of statistical techniques that are robust to outliers and heavy tails (in contrast to the more classical approaches like averages and Pearson correlation).
Spearman correlation	A popular ordinal correlation.
Test statistic	A number that is computed from data that can be used to gauge the accuracy of the null hypothesis. If the test statistic falls within expected the range, then the null hypothesis is a plausible explanation for the data. An extreme enough test statistic allows us to reject the null hypothesis.

T-test   Given two collections of numbers (say sales revenue from employees at two different offices), this is a statistical test of whether those numbers come from underlying distributions with the same mean.

Weibull distribution   A probability distribution used to model the lifetime of things that display either infant mortality or wearing out. The exponential distribution is a special case of the Weibull distribution where neither infant mortality nor wear-out comes into play.

5

Machine Learning

If data scientists are known for anything, it is their ability to deploy machine learning models to solve problems. This chapter will cover what machine learning is, what you can do with it, and the key things to understand in order to get the most out of it.

Machine learning basically means any method of using the computer to find patterns in data – the term "pattern recognition" is often used synonymously. Usually it is looking for patterns that make some sort of prediction, like using somebody's history on a website to guess which ad they will click, using a patient's blood results to guess whether they have cancer, or using measurements taken on an assembly line to assess whether a component is likely to work if you put it into a final product.

The previous chapter talked about methods for using data to create and test narratives that humans can understand. This chapter takes the opposite approach: the primary goal of machine learning is to create software programs that make correct decisions with minimal human involvement (although human intuition can be crucial in crafting and vetting the models).

Machine learning models often make worse decisions than human domain experts who are familiar with the prior data – sometimes dramatically so. But they make those decisions in fractions of a second, consistently, and at scale. Any time users are presented with a recommendation engine, or software behaves differently to anticipate a somebody's actions, or the computer flags something as being worthy of human attention, the logic making those decisions is liable to be a machine learning model.

Data Science: The Executive Summary - A Technical Book for Non-Technical Professionals,
First Edition. Field Cady.
© 2021 John Wiley & Sons, Inc. Published 2021 by John Wiley & Sons, Inc.

5.1   Supervised Learning, Unsupervised Learning, and Binary Classifiers

Machine learning falls into two broad categories called "supervised learning" and "unsupervised learning." Both of them are based on finding patterns in historical data.

In supervised learning we have something specific (often called the "target" variable) that we are trying to predict about the data, and we know what the right predictions are for our historical data. The goal is specifically to find patterns that can be used to predict the target variable for other data in the future, when we won't know what the right answer is. This is by far the most high-profile, clearly useful application of machine learning.

In unsupervised learning there is no specific target variable that we are trying to predict. Instead the goal is to identify latent structures in the data, like the datapoints tending to fall into several natural clusters. Often information like that is not an end in itself, but gets used as input for supervised learning.

The simplest and most important type of supervised learning is the binary classifier, where the target variable we are trying to predict is a yes/no label, typically thought of as 1 or 0. Typically the labeled data is divided into "training data" and "testing data." The algorithm learns how to give the labels by pattern-matching to the training data, and we measure how well it performs by seeing the labels it gives to the testing data. The train/test split is to make sure that the patterns we found in the training data could be applied fruitfully to the other, testing data.

A wealth of business problems fit within this paradigm. For example, you can us a classifier to guess:

- Whether or not a patient has cancer, based on lab results
- Whether a visitor to a website will click an ad based on their behavior on the site
- Whether a machine that is running will fail within the next 30 days

The last of these starts to illustrate one of the subtle, business-related aspects of using a binary classifier. In many cases there is no mathematically "right" answer to what we should consider a yes/no – it's a business judgment. The choice of 30 days as a cutoff is arbitrary – in a real situation you would probably tune it based on the time scale required to fix a problem that arises.

This book will focus on binary classifiers as the poster child for supervised learning. There are three reasons for this:

- The most interesting applications tend to be binary
- Almost all of the concerns that are relevant to other supervised learning are true of binary classifiers

- Many other applications can be created using binary classifiers as building blocks. For example, if you want to label somebody with a US state that can be done with 50 binary classifiers, each of which distinguishes between a particular state and "other"

In some cases the label coming out of a binary classifier is literally a 1 or 0, but usually it is a decimal between 1 and 0, which indicates the level of confidence that the correct answer is a yes.

5.1.1 Reality Check: Getting Labeled Data and Assuming Independence

Pretty much all of the theoretical infrastructure of supervised machine learning and artificial intelligence rests on three key assumptions that are rarely 100% true:

- The target variables for your training data are known with certainty and are truly what you are trying to predict.
- All of your datapoints, both the labeled data you have and the future data you will be making predictions for, are independent of each other. Separate flips of a coin, distinct rolls of a dice, etc.
- Future data will have all the same statistical properties as the labeled data you are training and testing on

Mathematicians summarize the last two bullets by saying that your datapoints are "i.i.d" – Independent and Identically Distributed. When all of these assumptions are true then machine learning largely becomes a purely technical problem that can be delegated to data scientists, but business sense is needed to understand when they break down and how it can be accounted for.

In many cases the target variables are known for certainty and it is obvious what they should be. A visitor to a website either clicks on an ad or does not, a patient either has a particular disease or they don't, etc. In many cases though there is no obvious choice. Say, for example, that we are an airline, and we want to see if it's possible to predict whether frequent fliers will churn so that we can know who to send coupons to in the hopes of keeping their business. How do you define "churn"? It could be that their business stops cold-turkey at some point in time, it could be a large year-over-year drop in business, it could be that they stop using their rewards card – there are infinitely many ways to parse it, and no matter what you choose many customers will fall into a gray area.

In cases such as this, a common practice is to cut the gray-area customers out from training your classifier, and to only train on the clear churners and the clear non-churners. The hope is that the resulting classifier will give ambiguous predictions – around 0.5 – for customers in the gray area. As for the definition of churn itself, this is a business question; all I can say is that you'll probably benefit from having very simple definitions whenever possible.

Datapoints being i.i.d is a much bigger issue, one which applies to almost all real-world classification problems. Getting labeled data that is truly representative of the data you will want to make predictions about in the future is usually impossible. Production models often get applied to situations they were not trained on, and for which data was not initially available. For example, you might train a model to predict user behavior on a portfolio of websites, but if a new site gets added to the portfolio you won't know how well the model will work there until the data starts coming in from it. Whenever a model gets applied to a novel situation, you should expect to take a hit in performance, and to have to re-train it as data comes in from the new situation.

Bear in mind that the future itself is a novel situation, because the world changes over time. Sometimes large, identifiable events will break a model in an instant, but there is also "model drift" as the patterns of the world evolve.

Perhaps the trickiest problem to address is subtle correlations within the data you have. Let's take the previous example of predicting whether a machine will fail within 30 days. Say we have data giving stats on the machine every day for multiple years, and we know when all the failures occurred. It might seem like the datapoint for a particular Monday is independent of the datapoint for the next Tuesday, because the data were gathered on separate days. But in fact they are correlated – chances are whatever was happening in the machine on Monday was also happening on Tuesday, so we probably have one underlying phenomenon (i.e. the state of the engine and whether a failure was imminent) masquerading as two datapoints. It sounds impressive to say that I have 20 examples of data when failure was imminent, until you realize they were just consecutive days leading up to the same failure event.

Sometimes correlations in the data can be partly addressed by assigning different weights to different datapoints. In general though there are not mathematical solutions to these problems – they are a guide for when you are creating machine learning models and a caveat to keep in mind when you are applying them.

5.1.2 Feature Extraction and the Limitations of Machine Learning

The most important caveat about machine learning models is that the computer does *not* have anything we could call an "understanding" or "common sense" about the data. Instead it is grasping to find numerical rules of thumb that correlate well with whatever it is trying to predict. Some of those rules of thumb work for causal reasons that a human can understand after the fact and reason about, some work for reasons that are convoluted, and some are just dumb luck; the computer can't tell the difference.

There is a constant tension in machine learning between wanting a model that captures the "real" dynamics of the world and one that has maximum accuracy. The former can be analyzed by humans to get insights, and is more likely to generalize to new situations, but it may leave money on the table by ignoring correlations that don't fall into a human-friendly narrative. The latter often has better performance statistics, but it may rely on incidental aspects of the world that can change without notice and it is liable to have nasty biases like racism and sexism.

This tradeoff will become a bigger deal in the chapter on artificial intelligence where we will encounter machine learning models, like deep neural networks, that are almost impossible to make sense of but nonetheless work extremely well. For the models we will discuss in this chapter though you can usually (though not always) have your cake and eat it too, by extracting numerical features from the data that do a good job of quantifying salient aspects of the world.

The reality is that machine learning models fit very crude formulas to the data. If those formulas are to perform well their inputs need to already be meaningful. This is one of the key areas of data science, but unfortunately I have very little to say about it because feature extraction is as varied as the problems to which data science is applied. If you read a book about machine learning – including this one – it will spend most of its time talking about specific classifiers and general methods to assess models, but most of the actual creativity and math goes into problem-specific feature extraction.

I mentioned that the main purpose of machine learning is to make software that makes the right judgment calls on its own, with human understanding (in the form of feature extraction) being simply a means to an end. That's only partly true. It is often possible to take a machine learning model that performs well, reverse-engineer it to find what patterns it is honing in on, and make business decisions based on those. For example, you could find that using a particular feature on a website is an indicator of people not coming back. Or perhaps machines will fail if they are operating under conditions X and were last repaired using component Y. In these cases, where models get dissected for actionable business insights, good feature extraction can be even more important.

5.1.3 Overfitting

The great bogeyman of machine learning is the fear that your model will learn patterns that work for the data you train it with, but that do not generalize. This can happen for the reasons discussed previously, where your data is not representative of what will be seen in the future. But it also happens with perfectly clean, independent datapoints, and in this case it is called "overfitting."

Recall that a machine learning model works by looking for patterns in the data that correlate with the target variables. Sometimes those are real-world patterns that generalize, and sometimes they are true just by coincidence. Say, for example, that you are trying to predict whether a drug will treat a disease based on a patient's DNA test. There are many gene variants in the human genome, and probably only a small number of patients in your data; there is a very good chance that all of those for whom the drug works will have *some* gene variant in common just by coincidence, and a machine learning model will latch onto that as a key indicator. In effect it is just memorizing the right answers for this particular dataset.

The more inputted features you have, the more likely it is that some of them will correlate with the target variable just by coincidence and your model will be overfitted. On the other hand if you have more labeled datapoints it is harder for a large-scale coincidence to occur. So having a large ratio of labeled datapoints to inputted features is the key to reducing overfitting. This is one of the reasons that feature extraction is important – it reduces the number of raw inputs.

We can't avoid overfitting completely, but what we can do is to estimate how well your model performs accounting for it. This is done via a process called "cross validation," where we divide our labeled data into "training data" and "test data." You tune a model to the training data, but you only evaluate its performance on the test data.

There are various ways to divide data into the testing and training sets, and I will discuss them in Section 5.1.4. But the key concept is that your model will always be somewhat overfitted, so you can only honestly evaluate its performance by seeing how well it does on the testing data.

5.1.4 Cross-Validation Strategies

Generally speaking the more training data you have the better a model will be, but the more testing data you have the more accurately you can measure the model's performance. The final model that runs in production will typically be trained on all available data for that reason, and we can hope that it will work a little bit better than the models we test during cross-validation.

The simplest type of train/test split is simply to randomly partition your labeled data between training and testing data, often with an 80% and 20% split, respectively.

A more rigorous variant is called "k-fold cross validation," where k is some number. The data is randomly partitioned into k sets of equal size. We use $k - 1$ of the partitions as our training data, and test on the remaining partition. You do this k different times, so that you train k slightly different models and each partition gets a turn as the testing data. Finally, you average your performance metrics across all of the k different models. This is shown graphically in Figure 5.1. The extreme version of this is "leave-one-out" cross-validation, where k is equal to the number of

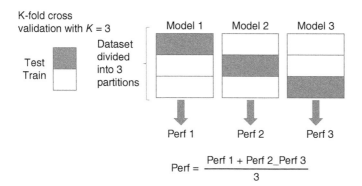

Figure 5.1 *K*-fold cross-validation breaks the dataset into *k* partitions. Each partition gets held out as the testing data, while the model is trained on the remaining $k - 1$ partitions. The average of all these performances indicates how good a model you were fitting.

datapoints you have and every testing dataset is only one point. This is impractical to do with all but the smallest datasets.

Sometimes the data comes in several different non-overlapping categories (men and women, machines purchased from one vendor versus another, etc.) called "strata." In these cases we may want to make sure that all the strata are appropriately represented in both the testing and training data. To do this we use "stratified sampling," where each stratum is divided into training/testing data, in the same proportions.

5.2 Measuring Performance

In most business applications of classification, you are looking for one class more so than the other. For example, you are looking for promising stocks in a large pool of potential duds. Or you are looking for patients who have cancer. The job of a classifier is to flag the interesting ones.

There are two aspects of how well a classifier performs: you want to flag the things you're looking for, but you also want to not flag the things you aren't looking for. Flag aggressively and you'll get a lot of false positives – potentially very dangerous if you're looking for promising stocks to invest in. Flag conservatively, and you'll be leaving out many things that should have been flagged – terrible if you're screening people for cancer. How to strike the balance between false positives and false negatives is a business question that can't be answered analytically.

For this chapter we will focus on two performance metrics that, together, give the full picture of how well a classifier performs:

- *True positive rate (TPR)*: Of all things that should be flagged by our classifier, this is the fraction that actually gets flagged. We want it to be high: 1.0 is perfect.
- *False positive rate (FPR)*: Of all things that should NOT be flagged, this is the fraction that still ends up getting flagged. We want it low: 0.0 is perfect.

I will give you a nice, graphical way to think of TPR and FPR that is the main way I think about classifiers in my own work.

But you could pick other metrics too – they're all equivalent. The other option you're most likely to see is "precision" and "recall." Precision is the same thing as the TPR – the fraction of all flagged results that actually should have been flagged. Recall measures your classifier's coverage – out of all the things that *should* be flagged it is the fraction that actually *gets* flagged.

5.2.1 Confusion Matrices

A common way to display performance metrics for a binary classifier is with a "confusion matrix." It is a 2×2 matrix displaying how many points in your testing data were placed in which category versus which category they should have been placed in. For example:

Correct label	Predicted = 0	Predicted = 1
0	35	4
1	1	10

In the confusion matrix earlier, the TPR would then be $10/(10+1) = 0.91$, and the FPR would be $4/(4+35) = 0.10$.

5.2.2 ROC Curves

If you treat the FPR as an x-coordinate and the TPR as the y-coordinate, then you can visualize a classifier's performance as a location in a two-dimensional box, shown in Figure 5.2. The upper-left corner $(0.0, 1.0)$ corresponds to a perfect classifier, flagging every relevant item with no false positives. The lower-left corner means flagging nothing, and the upper-right means flagging everything. If your classifier is below the $y = x$ line, then it's worse than useless; an irrelevant item is more likely to be flagged than one that's actually relevant.

My discussion so far has talked about classifiers in a binary way; they will label something as (say) fraud or non-fraud. But very few classifiers are truly binary; most of them output some sort of a score, and it's up to data scientists to pick a cutoff for what gets flagged. This means that a single classifier is really a whole

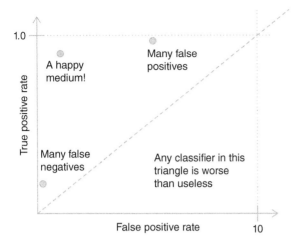

Figure 5.2 The performance of a classifier can't really be boiled down to a single number, because false positives and false negatives are two separate phenomena. Which one is worse depends on the business application. This chart puts the false positive rate – the fraction of 0s that are incorrectly flagged as 1 – on the x-axis. The y-axis is the true positive rate – that fraction of 1s that are correctly flagged as such. If a classifier is overly generous in labeling data with 1 it will be in the upper right. Overly stingy and it will be in the lower left.

Figure 5.3 The ROC curve plots the true/false positive rate for a classifier over the whole range of its thresholds. A good classifier will have a high arc over the FPR = TPR line.

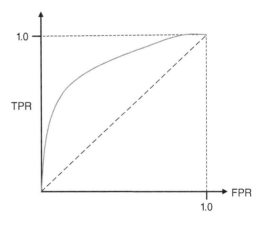

family of classifiers, corresponding to where we pick the cutoff. Each of these cut-offs corresponds to a different location in our 2d box, and together they trace out what's called an ROC curve, an example of which is shown in Figure 5.3.

Imagine you start with an insanely high classification threshold, so high that nothing actually gets flagged. This means you are starting in the lower-left of the box, at (0, 0). As you loosen your criteria and start flagging a few results, your location shifts. Hopefully the first results you start flagging are all dead-ringers and you get very few false positives, i.e. your curve slopes sharply up from the origin. When you get to the knee of the curve you have flagged all the low-hanging fruit, and hopefully not incurred many false positives. If you continue to loosen your criteria, you will start to correctly flag the stragglers, but you will also flag a lot of false positives. Eventually, everything will get flagged and you will be at (1, 1).

If we are trying to understand the quality of the underlying score-based classifier, then it's not fair to judge it by the performance at any single threshold. You want to judge it by the entire ROC curve – a sharp "knee" jutting into the upper-left corner is the signature of a strong classifier.

5.2.3 Area Under the ROC Curve

This holistic, look-at-the-whole-roc-curve viewpoint doesn't absolve us from sometimes having to boil performance down into a single number. Sometimes you'll need a numerical criteria for, say, declaring that one configuration for your classifier is better than another.

The standard way to score an entire ROC curve is to calculate the area under the curve (AUC); a good classifier will mostly fill the square and have an AUC near 1.0, but a poor one will be close to 0.5. The AUC is a good way to score the underlying classifier, since it makes no reference to where we would draw a classification cutoff.

Another advantage of the AUC is that it has a very clear, real-world meaning. If you randomly select a datapoint from your data that is a 1, and another that is a 0, then the AUC is the probability that the former will have a larger score than the latter.

In real situations the AUC is often used to decide which of several machine learning models to use, because it is seen as reflecting how well the model has truly picked up on signal in the data. If you are dissecting the model for business insights you can stop there. If you are using it in production you then have a separate question of where best to set the classification thresholds on the model.

5.2.4 Selecting Classification Cutoffs

Intuitively we want to set our thresholds so that our classifier is near the "knee" of the curve. Maybe business considerations will nudge us to one part or another of the knee, depending on how we value precision versus recall, but there are also mathematically elegant ways to do it. I'll discuss two of the most common ones in this section.

The first is to look at where the ROC curve intersects the line $y = 1 - x$, as in Figure 5.4.

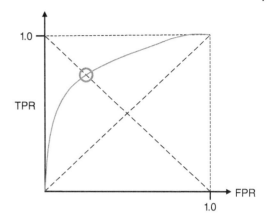

Figure 5.4 For this cutoff the fraction of all 0s that get incorrectly flagged as 1s is equal to the fraction of all 1s that get incorrectly flagged as 0s.

This means that the fraction of all hits that end up getting flagged is equal to the fraction of all non-hits that *don't* get flagged: we have the same accuracy on the hits and the non-hits. A different cutoff would make me do better on hits but worse on non-hits, or vice versa. All other things being equal, this is the cutoff that I use, partly because I can truthfully answer the question of "how accurate is your classifier?" with a single number.

The second approach is to look at where the ROC curve has a 45° slope, i.e. where it runs parallel to the line $y = x$ as in Figure 5.5. This is an inflection point where: below this threshold, relaxing your classifier boosts the flagging probability of a hit more so than a non-hit. Above this threshold, relaxing your classifier will boost the flagging probability for non-hits more so than for hits. It's effectively like saying that an epsilon increase in TPR is worth the cost of an epsilon increase in FPR, but no more than that.

This second approach is also useful because it generalizes. You could instead decide that a tiny increase in TPR is worth three times the increase in FPR, because it's that much more important to you to find extra hits. Personally I've never had occasion to go that far into the weeds, but you should be aware it's possible in case the need arises.

5.2.5 Other Performance Metrics

The AUC is the right metric to use when you're trying to gauge the holistic performance of a classifier that gives out continuous scores. But if you're using the classifier as a basis for making decisions, the underlying classifier score is only worth as much as the best single classifier you can make out of it. So what you might do here is to pick a "reasonable" threshold (by the definition of your choice) for your classifier, and then evaluate the performance of that truly binary classifier.

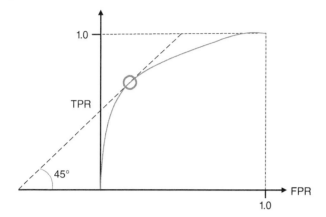

Figure 5.5 For this cutoff a small change in your classification threshold will increase the true positive and false positive rates by the same amount.

The classical way to judge a binary classifier is called the F_1 score. It is the harmonic mean of the classifier's precision with its recall, defined by

$$F_1 = \frac{1}{\left(\frac{1}{\text{Precision}} + \frac{1}{\text{Recall}}\right)/2}$$

$$= \frac{2 \times \text{Precision} \times \text{Recall}}{\text{Precision} + \text{Recall}}$$

The F_1 score will be 1.0 for a perfect classifier and 0.0 in the worst case. It's worth noting though that there is nothing magical about using the harmonic mean of precision and recall, and sometimes you will see the geometric mean used to compute the G-score:

$$G = \sqrt{\text{Precision} * \text{Recall}}$$

You might be tempted to use the arithmetic mean (precision + recall)/2, but that would have the unpleasant effect that flagging everything (or not flagging anything at all) would give a score better than zero.

5.2.6 Lift Curves

Some people prefer what's called a lift curve, rather than an ROC curve. It captures equivalent information, namely, how the performance of a classifier varies as you adjust the classification threshold, but displays it in a different way. The lift reach curve is based on the following notions:

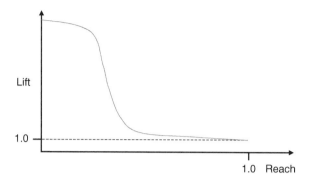

Figure 5.6 In a lift curve the *x*-axis (the "reach") is the fraction of all data that gets labeled as a 1. Lift is the precision (fraction of things getting labeled as 1 that are actually hits) divided by the actual frequency of hits. The stingier we are about labeling data as 1 the more accurate we expect those labels to be. When all data gets flagged our precision is simply the fraction of 1s in the population.

- Reach is the fraction of all points that get flagged.
- Lift is the fraction of all points you flag that are hits, divided by the fraction of hits in the overall population. A lift of 1.0 means that you are just flagging randomly, and anything above that is positive performance.

The reach is plotted along the *x*-axis, and the lift along the *y* axis. Typically the lift will start off high, and then it decays to 1.0 as reach approaches 1. A good plot might look like Figure 5.6.

5.3 Advanced Material: Important Classifiers

The world is full of different classification algorithms. This section will go over some of the most useful and important ones.

5.3.1 Decision Trees

A decision tree is conceptually one of the simplest classifiers available. Using a decision tree to classify a datapoint is the equivalent of following a basic flowchart. It consists of a tree structure like the one in Figure 5.7.

Every node in the tree asks a question about one feature of a datapoint. If the feature is numerical, the node asks whether it is above or below a threshold, and there are child nodes for "yes" and "no." If the feature is categorical, typically there will be a different child node for every value it can take. A leaf node in the tree will

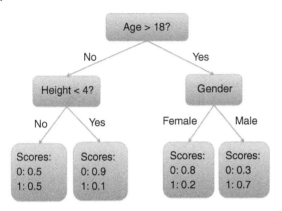

Figure 5.7 A decision tree classifier is somewhat like a flow chart. Every node puts a threshold on some variable – if you are below that threshold go to the left and otherwise go to the right (for categorical variables, like gender, it's a yes-no question rather than a threshold). At the bottom of the flow chart you have your prediction.

be the score that is assigned to the point being classified (or several scores, one for each possible thing the point could be flagged as). It doesn't get much simpler than that.

Using a decision tree is conceptually quite straightforward, but training one is another matter entirely. In general, finding the optimal decision tree for your training data is computationally intractable, so in practice you train the tree with a series of heuristics and hope that the result is close to the best tree possible. Generally the algorithm is something along these lines:

1. Given your training data X, find the single feature (and cutoff for that feature, if it's numerical) that best partitions your data into classes.
2. There are a variety of ways to quantify how good a partition is. The most common ones are the "information gain" and the "Gini impurity." I won't delve into their precise meanings here.
3. This single best feature/cutoff becomes the root of your decision tree. Partition X up according to this node.
4. Recursively train each of the child nodes on its partition of the data.
5. The recursion stops when either (i) all of the datapoints in your partition have the same label, (ii) the recursion has gone to a pre-determined maximum depth, or (iii) there are so few datapoints in the partition that we don't want to split them up any further. At that point, the scores stored in this node will just be the breakdowns of the labels in the partition.

In step 5 the maximum tree depth and the minimum datapoints to split the partition are parameters that we can tune to ensure that we don't overfit. If they are too lax the training will recurse too deeply and give definitive 0/1 answers that memorize the training data, rather than floating point answers that generalize better. Note that in many numerical libraries the default values of these parameters are quite lax.

Decision trees are very easy to understand, so it's perhaps a bit surprising that they are often difficult to tease real-world insights out of. Looking at the top few layers is certainly interesting, and suggests what some of the more important features are. But it's not always clear what the features and their cutoffs mean for the real world, unless you want to wade into the deep waters of dissecting the Gini impurities from the training stage. Even if you do this, there is still a very real risk that the same feature will weigh toward hits at one node of the tree, and toward non-hits at another node. What the heck does that mean?

Personally, I don't use decision trees much for serious work. However, they are extremely useful for their human-readability – this is especially handy if you're working with people who don't know machine learning and are wary of black boxes – and the rapidity with which they can do classifications. Above all, decision trees are useful as a building block for constructing Random Forest classifiers, which I'll discuss in Section 5.3.2.

5.3.2 Random Forests

If I were stuck on a desert island and could only take one classifier with me, it would be the random forest. They are consistently one of the most accurate, robust classifiers out there, legendary for taking in datasets with a dizzying number of features, none of which are very informative and none of which have been cleaned, and somehow churning out results that beat the pants off of anything else.

The basic idea is almost too simple. A random forest is a collection of decision trees, each of which is trained on a random subset of the training data, and only allowed to use some random subset of the features. There is no coordination in the randomization – a particular datapoint or feature could randomly get plugged into all the trees, none of the trees, or anything in between. The final classification score for a point is the average of the scores from all the trees (or sometimes you treat the decision trees as binary classifiers, and report the fraction of all of them that votes a certain way).

The hope is that the different trees will pick up on different salient patterns, and each one will only give confident guesses when its pattern is present. That way, when it comes time to classify a point, several of the trees will classify it correctly and strongly while the other trees give answers that are on-the-fence, meaning the overall classifier slouches toward the correct answer.

The individual trees in a random forest are subject to overfitting, but they tend to be randomly overfitted in different ways. These largely cancel each other out, yielding a robust classifier.

The problem with random forests is that they're impossible to make real business sense of. The whole point of a classifier like this is that it is too complex for human comprehension, and its performance is an averaged-out thing.

The one thing that you can do with a random forest is to get a "feature importance" score for any feature in the dataset. These scores are opaque, and impossible to ascribe a specific real-world meaning to. The importance of the kth feature is calculated by randomly swapping the kth feature between the points in the training data, and then looking at how much randomizing this feature hurts performance (there is a little bit of extra logic to make sure that no randomized datapoint is fed into a tree that was trained on its non-randomized version). In practice you can often take this list of features and, with a little bit of old-fashioned data analysis, figure out compelling real-world interpretations of what they mean. But the random forest itself tells you nothing.

5.3.3 Ensemble Classifiers

Random forests are the best-known example of what are called "ensemble classifiers," where a wide range of classifiers (decision trees in that case) are trained under randomly different conditions (in that case, random selections of datapoints and features) and their results are aggregated. Intuitively the idea is that if every classifier is at least marginally good, and the different classifiers are not very correlated with each other, then the ensemble as a whole will very reliably slouch toward the right classification. Basically it's using raw computational power in lieu of domain knowledge or mathematical sophistication, relying on the power of the law of large numbers.

5.3.4 Support Vector Machines

I'll be honest: I personally hate support vector machines (SVMs). They're one of the most famous machine learning classifiers out there, so it's important that you be familiar with them, but I have several gripes. First off, they make a very strong assumption about the data called linear separability. Oftentimes that assumption is wrong, and occasionally it's right in mathematically perverse ways. There are sometimes hacks that work around this assumption, but there's no principal behind them, and no a priori way of knowing which (if any) hack will work in a particular situation. SVMs are also one of the few classifiers that are fundamentally binary; they don't give continuous-valued "scores" that can be used to assess how confident the classifier is. This makes them annoying if you're looking for business insights, and unusable if you need to have the notion of a "gray area."

That said, they're popular for a reason. They are intuitively simple, mathematically elegant, and trivial to use. Plus, those un-principled hacks I mentioned earlier can be incredibly powerful if you pick the right one.

Figure 5.8 Support vector machines look for a hyperplane that divides your training data into the 0s and the 1s.

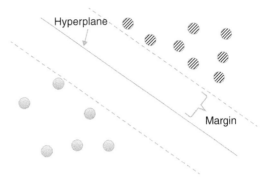

The key idea of an SVM is illustrated in Figure 5.8. Essentially you view every datapoint as a point in d-dimensional space, and then look for a hyperplane that separates the two classes. The assumption that there actually is such a hyperplane is called linear separability. Training the SVM involves finding the hyperplane that (i) separates the datasets and (ii) is "in the middle" of the gap between the two classes. Specifically, the "margin" of a hyperplane is min (its distance to the nearest point in class A, its distance to the nearest point in class B), and you pick the hyperplane that maximizes the margin.

Mathematically, the hyperplane is specified by the equation

$$f(x) = w \cdot x + b = 0$$

where w is a vector perpendicular to the hyperplane and b measures how far offset it is from the origin. To classify a point x, simply calculate $f(x)$ and see whether it is positive or negative. Training the classifier consists of finding the w and b that separates the dataset while having the largest margin.

This version is called "hard margin" SVM. However, in practice there often is no hyperplane that completely separates the two classes in the training data. Intuitively what we want to do is find the best hyperplane that *almost* separates the data, by penalizing any points that are on the wrong side of hyperplane. This is done using "soft margin" SVM.

The other killer problem with SVM is if you have as many features as datapoints. In this case there is guaranteed to be a separating hyperplane, regardless of how the points are labeled. This is one of the curses of working in high-dimensional space. You can do dimensionality reduction (which I discuss in a later chapter) as a pre-processing step, but if you just plug high-dimensional data into an SVM, it is almost guaranteed to be grotesquely overfitted.

The most notorious problem with a plain SVM is the linear separability assumption. An SVM will fail utterly on a dataset like Figure 5.9 because there is no line between the two classes of points. The pattern is clear if you just look at it – one class is near the origin, and the other is far from it – but an SVM can't tell. The

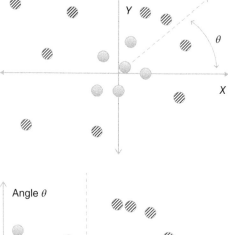

Figure 5.9 The key weakness of support vector machines is that often there is no hyperplane. In this care the pattern is clear enough – points close to the origin are solid further away they are shaded – but there is no plane that separates them.

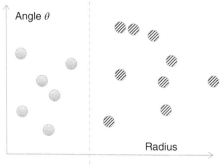

Figure 5.10 Sometimes you can fix the linear separability problem by mapping the data into a different space where there is a hyperplane that divides the two classes. The trick is to find what that mapping is!

solution to this problem is a very powerful generalization of SVMs called "kernel SVM." The idea of Kernel SVM is to first map our points into some other space in which decision boundary is linear, and do SVM there. For Figure 5.9, if we plot the distance from the origin on the x-axis and the angle θ on the y-axis, we get Figure 5.10, which is linearly separable. In general, kernel SVM requires finding some function ϕ that maps our points in d-dimensional space to points in some n-dimensional space. In the example I gave n and d were both 2, but in practice we usually want n to be larger than d, to increase the chances of linear separability. If you can find ϕ then you're golden.

Now here's the key point computationally: you never need to find ϕ itself. When you crank through the math, it turns out that whenever you calculate $\phi(x)$, it is always part of a larger expression called the kernel function:

$$k(x, y) = \phi(x) \cdot \phi(y)$$

The kernel function takes two points in the original space and gives their dot product in the mapped space. This means that the mapping function ϕ is just an abstraction – we never need to calculate it directly, and can instead just focus on k. And in many cases it turns out that calculating k directly is much, much easier than calculating any $\phi(x)$ intermediates. It is often the case that ϕ is an intricate mapping into a massively high-dimensional space, or even an infinite-dimensional space, but the expression for k reduces to some simple, tractable function that is non-linear. Using only the Kernel function in this way is called the "kernel trick," and it ends up applying to areas outside of SVMs.

Not every function that takes in two vectors is a valid kernel, but an awful lot of them are. Some of the most popular, which are typically built in to libraries, are:

- *Polynomial kernel:* $k(x, y) = (x \cdot y + c)^n$
- *Gaussian kernel:* $k(x, y) = \mathrm{Exp}[-\gamma |x - y|^2]$
- *Sigmoid:* $k(x, y) = \tan h(x \cdot y + r)$

Most Kernel SVM frameworks have these kernels available out-of-the-box, and may also let used define their own kernel functions as well. Be extra cautious about the latter approach though, because it is usually the user's job to make sure that whatever they define meets the mathematical criteria for being a kernel.

5.3.5 Logistic Regression

Logistic regression is a great general-purpose classifier, striking an excellent balance between accurate classifications and real-world interpretability. I think of it as kind of a non-binary version of SVM, one that scores points with probabilities based on how far they are from the hyperplane, rather than using that hyperplane as a definitive cutoff. If the training data is almost linearly separable, then all points that aren't near the hyperplane with get a confident prediction near 0 or 1. But if the two classes bleed over the hyperplane, a lot the predictions will be more muted, and only points far from the hyperplane will get confident scores.

In logistic regression the score for a point will be

$$p(x) = \frac{1}{1 + \mathrm{Exp}[w \cdot x + b]}$$

Note that $\mathrm{Exp}[w \cdot x + b]$ uses the same $f(x)$ we saw in SVM, where w is a vector that gives weights to each feature and b is a real-valued offset. With SVM we look at whether $f(x)$ is positive or negative, but in this case we plug it into the so-called sigmoid function:

$$\sigma(z) = \frac{1}{1 + \mathrm{Exp}[z]}$$

As with SVM, we have a dividing hyperplane defined by $w \cdot x + b = 0$. In SVM that hyperplane is the binary decision boundary, but in this case it is the hyperplane along which $p(x) = \frac{1}{2}$.

The sigmoid function shows up a few places in machine learning so it makes sense to dwell on it a bit. If you plot out $\sigma(x)$, it looks like Figure 5.11. You can see

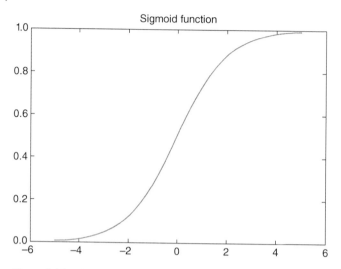

Figure 5.11 The Sigmoid function shows up many places in machine learning. Effectively it converts a confidence weight (which can take any value) to a number between 0 and 1. Strongly negative weights make it close to 0, large positive weights are close to 1, and a weight of 0 give 0.5.

that $\sigma(0)$ is 0.5. As the argument blows up to infinity, it approaches 1.0, and as it goes to negative infinity it goes 0.0. Intuitively, this makes it a great way to take "confidence weights" and cram them down into the interval (0, 1.0) where they can be treated as probabilities. The sigmoid function also has a lot of convenient mathematical properties that make it easy to work with. We will see it again in the section on neural networks.

Pulling real-world meaning out of a trained logistic regression model is easy:

- If the kth component of w is large and positive, then the kth feature being big suggests that the correct label is 1.
- If the kth component of w is large and negative, then the kth feature being big suggests that the correct label is 0.
- The larger the elements of w are in general, the tighter our decision boundary and the more closely we approach an SVM.

Note though that in order for this to be meaningful, you must make sure your data is all set to the same scale before training; if the most important features also happens to be the largest numbers, then its coefficient would be misleadingly small.

Another perk of logistic regression is that it's extremely efficient to store and use. The entire model consists of just $d + 1$ floating point numbers, for the d components of the weights vector and the offset b. Performing a classification requires just d multiplication operations, d addition operations, and one computation of a sigmoid function.

5.3.6 Lasso Regression

Lasso regression is a variant of logistic regression. One of the problems with logistic regression is that you can have many different features all with modest weights, instead of a few clearly meaningful features with large weights. This makes it harder to extract real-world meaning from the model. It is also an insidious form of overfitting, which is begging to have the model generalize poorly.

In lasso regression $p(x)$ has the same functional form of $\sigma(w \cdot x + b)$, where we assign every feature a weight and then plug their weighted sum into a logistic function. However, we train it in a different way that penalizes modest-sized weights. Depending on how harsh the penalty is (this is a parameter that can be tuned), this tends to yield a model where the majority of features are assigned a weight close to zero while a small handful of features carry essentially all the classification weight.

The numerical algorithm that finds the optimal weights generally doesn't use heuristics or anything; it's just cold numerical trudging. However, as an aid to human intuition, I like to think of some examples of heuristics that the solver might, in effect, employ:

- If features i and j have large weights, but they usually cancel each other out when classifying a point, set both their weights to zero.
- If features i and j are highly correlated you can reduce the weight for one while increasing the weight for the other and keeping predictions more-or-less the same.

5.3.7 Naive Bayes

Bayesian statistics is one of the biggest, most interesting, and most mathematically sophisticated areas of machine learning. However, most of that is in the context of Bayesian networks which are a deep, highly sophisticated family of models that tend to be carefully crafted for the system you are hoping to describe. Data scientists are more likely to use a drastically simplified version called Naive Bayes.

I talk in more detail about Bayesian statistics in the chapter on statistics. Briefly though, a Bayesian classifier operates on the following intuition: you start off with some initial confidence in the labels 0 and 1 (assume it's a binary classification problem). When a new piece of information becomes available, you adjust your confidence levels depending on how likely that information is conditioned on each label. When you've gone through all available information, your final confidence levels are seen as the probabilities of the labels 0 and 1.

Ok, now let's get more technical. During the training phase, a naïve Bayesian classifier learns two things from the training data:

- How common every label is in the whole training data
- For every feature X_i, its probability distribution *when the label is 0*
- For every feature X_i, its probability distribution *when the label is 1*

Those last two are called the conditional probabilities, and they are written as

$$\Pr(X_i = x_i \mid Y = 0)$$

$$\Pr(X_i = x_i \mid Y = 1)$$

When it comes time to classify a point $X = (X_1, X_2, \ldots, X_d)$, the classifier starts off with confidences

$$\Pr(Y = 0) = \text{fraction of the training data with } Y = 0$$

$$\Pr(Y = 1) = \text{fraction of the training data with } Y = 1$$

Then for each feature X_i in the data, let x_i be the value it actually had. We then update our confidences to

$$\Pr(Y = 0) \leftarrow \Pr(Y = 0) * \Pr(X_i = x_i \mid Y = 0) * \gamma$$

$$\Pr(Y = 1) \leftarrow \Pr(Y = 1) * \Pr(X_i = x_i \mid Y = 1) * \gamma$$

where we set γ so that the confidences add up to 1.

There are a lot of things here that need fleshing out if you're implementing a naïve Bayes classifier. For example, we need to assume some functional form for $\Pr(X_i = x_i \mid Y = 0)$, like a normal distribution or something, in order to fit it during the training stage (there is nothing to flesh out though if everything is modelled as a 0/1 coin flip, which is part of why binary features are very popular in Bayesian contexts). We also need to be equipped to deal with over-fitting there.

But the biggest problem is that we are treating each X_i as being independent of every other X_j. In a real situation our data might be such that X_5 is highly correlated with X_4. In that case we really shouldn't adjust our confidences when we get to X_5, since X_4 had already accounted for it. Basically we are double-counting whatever underlying thing causes X_5 and X_4 to be correlated. So it's perhaps surprising that Naïve Bayes tend to be a very powerful classifier in practice. The way I think of it is this: if X_4 is truly a powerful predictive variable and we are double-counting it, then this might make us overconfident but we are usually overconfident in the right direction.

5.3.8 Neural Nets

Neural nets used to be the black sheep of classifiers, but they have enjoyed a renaissance in recent years, especially the sophisticated variants collectively known as "deep learning." Neural nets are as massive an area as Bayesian networks, and many people make careers out of them. However, basic neural nets are standard tools that you see in machine learning. They are simple to use, fairly effective as classifiers, and useful for teasing interesting features out of a dataset.

Neural nets were inspired by the workings of the human brain, but now that we know more about how biological circuits work, it's clear that that analogy is bunk. Really sophisticated deep learning is at the point where it can be compared with some parts of real brains (or maybe we just don't know enough about the brain yet to see how much they fall short), but anything short of that should be thought of as just another classifier.

The simplest neural network is the perceptron. A perceptron is a network of "neurons," each of which takes in multiple inputs and produces a single output. An example is shown in this Figure 5.12.

The labeled nodes correspond to either the input variables to a classification or a range of output variables. The other nodes are neurons. The neurons in the first layer take all of the raw features as inputs. Their outputs are fed as inputs to the second layer, and so on. Ultimately, the outputs of the final layer constitute the output of your program. All layers of neurons before the last one are called "hidden" layers (there is one in this figure the way I've drawn it). Unlike other classifiers, neural networks very organically produce an arbitrary number of different outputs, one for each neuron in the final layer. In this case there are three outputs. In general you can use neural nets for tasks other than classification, and treat the outputs as a general-purpose numerical vector. In classification tasks though, we typically look at the ith output as the score for the ith category that a point can be classified as.

The key part of a neural net is how each neuron determines its output from its various inputs. This is called the "activation function," and there are a number of

Figure 5.12 A perceptron is a neural network with a single hidden layer.

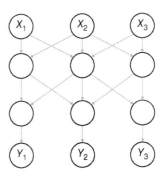

options you can pick from. The one I've seen the most is our old friend the sigmoid function. If we let i indicate some particular neuron, and j ranges over its inputs, then:

$$\text{Activation}_i = \sigma\left(b_i + \sum_j w_{ij} * \text{Input}_j\right)$$

In effect each neuron in the system is its own little logistic regression function, operating on the inputs to that neuron. A neural network with no hidden layers is, in fact, just a collection of logistic regressors.

For me, neural networks are not a tool I use a lot. Simple ones like the perceptron don't perform particularly well, and using the more complicated ones is an extremely specialized area. I'm more of an ensemble classifier guy, trusting in the law of large numbers rather than the voodoo of deep learning. But that's just me. Neural nets are a hot area, and they are solving some very impressive problems. There's a good chance they will become a much larger, more standard tool in the data science toolkit.

5.4 Structure of the Data: Unsupervised Learning

This section is about techniques for studying the latent structure of your data, in situations where we don't know a priori what it should look like. They are often called "unsupervised" learning because, unlike classification and regression, there is no externally known ground truth that we are hoping to match. There are two primary ways of studying a dataset's structure: clustering and dimensionality reduction.

Clustering is an attempt to group the datapoints into distinct "clusters." Typically this is done in the hopes that the different clusters correspond to different underlying phenomena. For example, if you plotted people's height on the x-axis and their weight on the y-axis, you would see two more-or-less clear blobs, corresponding to men and women. An alien who knew nothing else about human biology might see this and hypothesize that we come in two distinct types.

In dimensionality reduction the goal isn't to look for distinct categories in the data. Instead, the idea is that the different fields are largely redundant, and we want to extract the real, underlying variability in the data. For example, in a high-resolution image adjacent pixels tend to have very similar values, and we want to boil this down into a much smaller number of summary numbers. Mathematically the idea is that your data is d-dimensional, but all of the points *actually* only lie on a k-dimensional subset of the space (with $k < d$), plus some d-dimensional noise. In 3D data your points could lie mostly just along a single line, or perhaps in a curved circle. Real situations of course are usually not so clean cut. It's more useful to think of k dimensions as capturing "most" of the variability in the data, and you can make k larger or smaller depending on how much of the information you want to re-produce.

A key practical difference between clustering and dimensionality reduction is that clustering is generally done in order to reveal the structure of the data, but dimensionality reduction is often motivated mostly by computational concerns. For example, if you're processing sound, image, or video files, d is likely to be tens of thousands. Processing your data then becomes a massive computational task, and there are fundamental problems that come with having more dimensions than datapoints (the "curse of dimensionality," which I'll discuss shortly). In these cases dimensionality reduction is a prerequisite for almost any analytics you might want to do, regardless of whether you're actually interested in the data's latent structure.

5.4.1   The Curse of Dimensionality

Geometry in high-dimensional spaces is weird. This is important, because a machine learning algorithm with d features operates on feature vectors that live in d-dimensional spaces. d can be quite large if your features are all of the pixel values in an image! In these cases the performance of these algorithms often starts to break down, and this decay is best understood as a pathology of high-dimensional geometry, the so-called curse of dimensionality.

The practical punchline to all of this is that if you want your algorithms to work well, you will usually need some way to cram your data down into a lower-dimensional space. There's no need to dwell too much on the curse of dimensionality – thinking about it can hurt the heads of us three-dimensional beings.

But if you're interested, I would like to give you at least a tiny taste of what goes on in high dimensions. Basically the problem is that in high dimensions different points get very far away from each other. In Figure 5.13 I randomly sampled 1000 points from a d-dimensional cube of length 1 on each side and made a histogram of how far all the points are form each other.

The results are shown in Figure 5.13. You can see that for $d = 500$ two points in the cube are almost always about the same distance from each other. If you did a similar simulation with spheres, you would see that almost all the mass of a high-dimensional sphere is in its crust.

Sound weird? Well yes, it is. That's why we reduce dimensions.

5.4.2   Principal Component Analysis and Factor Analysis

The grand-daddy of dimensionality reduction algorithms is without question principal component analysis, or PCA.

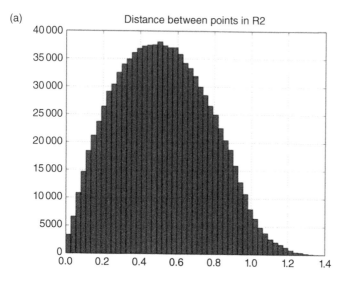

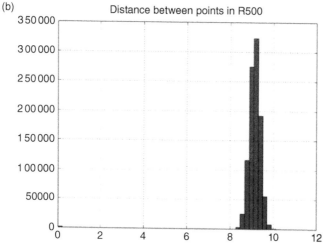

Figure 5.13 The "curse of dimensionality" describes how high-dimensional spaces have a number of geometric properties that seem quite bizarre relative to two- and three-dimensional space. In (a) we take 500 random pairs of points from a two-dimensional square (length 1.0 on a side) and make a histogram of the distances between them. You can see that there is a wide distribution – some pairs are far apart, some are close together, and most are midway. In (b) we do the same thing except that points are taken from a 500-dimensional hypercube. Almost all pairs of points are about the same distance apart. Properties like this can confound many machine learning models and other analytical techniques, so finding ways to reduce the dimensionality is often an important part of working with data that contains many numerical fields.

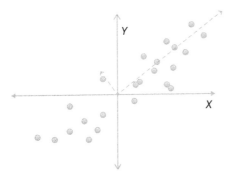

Figure 5.14 If many fields in your data move in lock-step then in a sense they are really just one degree of freedom. Geometrically this looks like your high-dimensional data lying mostly on a lower-dimensional subspace. Principal component analysis capitalizes on this, assuming that you data is much more spread out in one direction than another. It identifies what the most highly variable directions are and finds how much of the variation in your data they account for.

Geometrically, PCA assumes that your data in d-dimensional space is "football shaped" – an ellipsoidal blob that is stretched out along some axes, narrow in others, and generally free of any massive outliers. An example of this is shown in Figure 5.14.

Intuitively the data is "really" one-dimensional, lying on the line $x = y$, but there is some random noise that slightly perturbs each point. Rather than giving the two features x and y for each point, you can get a good approximation with just the single feature $x + y$. There are two ways to look at this:

- Intuitively, it seems like $x + y$ might be the real-world feature underlying the data, while x and y are just what we measured. By using $x + y$ we are extracting a feature that is more meaningful than any of our actual raw features.
- Technically, it is more computationally efficient to process one number rather than two. If you wanted, in this case, you could estimate x and y pretty accurately if you only knew $x + y$. Numerically, using one number rather than two lets us shy away from the curse of dimensionality.

PCA is way of (i) identifying the "correct" features like $x + y$ that capture most of the structure of the dataset and (ii) extracting these features from the raw datapoints.

To be a little more technical, PCA takes in a collection of d-dimensional vectors and finds a collection of d "principal component" vectors of length 1, called $p_1, p_2, \ldots$ and p_d. A point x in the data can be expressed as

$$x = a_1 p_1 + a_2 p_2 + \cdots + a_d p_d$$

However, the p_i is chosen so that generally a_1 is much larger than the other a_i, a_2 is larger than a_3 and above, etc. So realistically the first few p_i often capture most of the variation in the dataset, and x is a linear combination of the first few p_i, plus some small correction terms. The ideal case for PCA is something where large swaths of features tend to be highly correlated, like a photograph where adjacent pixels are likely to have similar brightness.

5.4.2.1 Scree Plots and Understanding Dimensionality

In our motivation for PCA we suggested that the dataset was "really" one-dimensional, and that one of the goals of PCA is to extract that "real" dimensionality of a dataset by seeing how many components it took to capture most of the dataset's structure. In reality it's rarely that clear. What you get instead is that the first few components are the most important, and then there is a gentle taper into uselessness as you go further out.

It is common to plot out the importance of the different principal components (technically the importance is measured as the fraction of a dataset's variance that the component accounts for) in what's called a "scree plot," like the one in Figure 5.15.

The x-axis tells us which component we are looking at – say, the nth. The y-axis is how much of the data's variability is along that component. For this plot we can see that after about six to seven dimensions the explained variability begins leveling off to a noisy baseline.

5.4.2.2 Factor Analysis

I should note that PCA is related to the statistical technique of factor analysis. Mathematically they're the same thing: you find a change of coordinates where most of the variance in your data exists in the first new coordinate, the second most in the second coordinate, etc. The divergent terminology is mostly an accident of

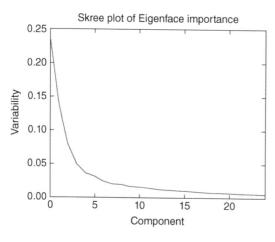

Figure 5.15 A Scree plot shows how much of a dataset's variability is accounted for by each principal component. In this plot by the time you get past six components the rest of them add very little.

the fact that they were developed independently and applied in very different ways to different fields.

In PCA the idea is generally dimensionality reduction; you find how many of these new coordinates are needed to capture most of your data's variance, and then you reduce your datapoints to just those few coordinates for purposes of something like plugging them into a classifier.

Factor analysis, on the other hand, is more about identifying causal "factors" that give rise to the observed data. A major historical example is in the study of intelligence. Tests of intelligence in many different areas (math, language, etc.) were seen to be correlated, suggesting that the same underlying factors could affect intelligence in many different areas. Researchers found that a single so-called "g-factor" accounts for about half of the variance between different intelligence tests. This book generally looks at things from a PCA perspective, but you should be aware that both viewpoints are useful.

5.4.2.3 Limitations of PCA

There are three big gotchas when using PCA:

- Your dimensions all need to be scaled to have comparable standard deviations. If you arbitrarily multiplied one of your features by a thousand (maybe by measuring a distance in millimeters rather than meters, which in principle does not change the actual content of your data), then PCA will consider that feature to contribute much more to the dataset's variance. If you are applying PCA to image data then there is a good chance this isn't a big deal, since all pixels are usually scaled the same. But if you are trying to perform PCA on demographic data, for example, you have to have somebody's income and their height measured to the same scale. Along with this limitation is the fact that PCA is very sensitive to outliers.
- PCA assumes that your data is linear. If the "real" shape of your dataset is that it's bent into an arc in high-dimensional space, it will get blurred into several principal components. PCA will still be useful for dimensionality reduction, but the components themselves are likely not to be very meaningful.
- If you are using PCA on images of faces or something like that, the key parts of the pictures need to be aligned with each other. PCA will be of drastically limited use if, for example, the eyes are covered by different pixels in different pictures. If your pictures are not aligned, doing automatic alignment is outside the skillset of most data scientists.

5.4.3 Clustering

Clustering is a bit of a dicier issue than using PCA. There are several reasons for this, but many of them boil down to the fact that it's clear what PCA is supposed to do, but we're usually not quite sure what we want out of clustering. There is no crisp analytical definition of "good" clusters; every candidate you might suggest has a variety of very reasonable objections you could raise. The only real metric is whether the clusters reflect some underlying natural segmentation, and that's very hard to assess: if you already know the natural segments, then why are you clustering?

To give you an idea of what we're up against, here are some of the questions to keep in the back of your mind

- What if our points fall on a continuum? This fundamentally baffles the notion of a cluster. How do I want to deal with that?
- Should clusters be able to overlap?
- Do I want my clusters to be nice little compact balls? Or would I allow something like a doughnut?

5.4.3.1 Real-World Assessment of Clusters

I will talk later about some analytical methods for assessing the quality of the clusters you found. One of the most useful of those is the Rand index, which allows you to compare your clusters against some known ground truth for what the clusters ought to be.

Usually though, you don't have any ground truth available, and the question of what exactly constitutes a good cluster is completely open-ended. In this case I recommend a battery of sanity checks. Some of my favorites include:

- For each cluster, calculate some summary statistics of it based on features that were NOT used as input to the clustering. If your clusters really correspond to distinct things in the real world, then they should differ in ways that they were not clustered on.
- Take some random samples from the different clusters and examine them by hand. Do the samples from different clusters seem plausibly different?
- If your data is high-dimensional, use PCA to project it down to just two dimensions and do a scatter plot. Do the clusters look distinct? This is especially useful if you were able to give a real-world interpretation to the principal components.
- Forget PCA. Pick two features from the data that you care about and do a scatterplot on just those two dimensions. Do the clusters look distinct?
- Try a different clustering algorithm. Do you get similar clusters?
- Redo the clustering on a random subset of your data. Do you still get similar clusters?

Another big thing to keep in mind is whether it is important to be able, in the future, to assign new points to one of the clusters we have found. In some algorithms there are crisp criteria for which cluster a point is in, and so it's easy to label new points that we didn't train on. In other algorithms though, a cluster is defined by the points contained in it, and assigning a new point to a cluster requires re-clustering the entire dataset (or doing some kind of a clever hack that's tantamount to that).

5.4.3.2 *k*-means Clustering

The k-means algorithm is one of the simplest techniques to understand, implement, and use. It starts with vectors in d-dimensional space, and the idea is to partition them into compact, non-overlapping clusters. I say again: the presumed clusters are compact (not loops, not super elongated, etc.) and not overlapping.

The classical algorithm to compute the clusters is quite simple. You start off with k cluster centers, then iteratively assign each data point to its closest cluster center and re-compute the new cluster centers. Here is the pseudo-code:

```
1. Start off with k initial cluster centers
2. Assign each point to the cluster center that it's
   closest to
3. For each cluster, re-compute its center as the
   average of all its assigned points
4. Repeat 2 and 3 until some stopping criterion is met.
```

There are clever ways to initialize the clusters if they are not present at the beginning, and to establish when they have become stable enough to stop, but otherwise the algorithm is very straightforward.

k-means clustering has an interesting property (sometimes a benefit) that when k is larger than the "actual" number of clusters in the data you will split a large "real" cluster into several computed ones. In this case k-means clustering is less of a way to identify clusters and more of a way to partition your dataset into regions. An example of how this can happen is shown in Figure 5.16, where we set $k = 3$ but there were "really" two clusters.

Situations like this can often be found using the silhouette score to find clusters that are not very distinct.

Figure 5.16 The "clusters" identified by *k*-means clustering are really just regions of space, especially if *k* is not equal to the number of "real" clusters in the data.

The results of k-means clustering are extremely easy to apply to new data; you simply compare a new data point with each of the cluster centers and assign it to the one that it's closest to.

An important caveat about k-means is that there is no guarantee about finding optimal clusters in any sense. For this reason it is common to re-start it several times with different, random initial cluster centers.

5.4.3.3 Advanced Material: Other Clustering Algorithms
Gaussian Mixture Models

A key feature of most clustering algorithms is that every point is assigned to a single cluster. But realistically many datasets contain a large gray area, and mixture models are a way to capture that.

You can think of Gaussian mixture models (GMMs) as a version of k-means that captures the notion of a gray area, and gives confidence levels whenever it assigns a point to a particular cluster.

Each cluster is modeled as a multi-variate Gaussian distribution, and the model is specified by giving

1. The number of clusters
2. The fraction of all datapoints that are in each cluster
3. Each cluster's mean and its d-by-d covariance matrix

When training a GMM, the computer keeps a running confidence level of how likely each point is to be in each cluster, and it never decides them definitively: the mean and standard deviation for a cluster will be influenced by every point in the training data, but in proportion to how likely they are to be in that cluster. When it comes time to cluster a new point, you get out a confidence level for every cluster in your model.

Mixture models have many of the same blessings and curses of k-means. They are simple to understand and implement. The computational costs are very light, and can be done in a distributed way. They provide clear, understandable output that can be easily used to cluster additional points in the future. On the other hand, they both assume more-or-less convex clusters, and they are both liable to fall into local minimums when training.

I should also note that GMMs are the most popular instance of a large family of mixture models. You could equally well have used something other than Gaussians as the models for your underlying clusters, or even had some clusters be Gaussians and others something else. Most mixture model libraries use Gaussians, but under the hood they are all trained with something called the EM algorithm, and it is agnostic to the distribution being modeled.

Agglomerative Clustering

Agglomerative clustering is a general class of algorithms sharing a common structure. We start off with a large number of small clusters, typically with each point being its own cluster. We then successively merge clusters together until they form a single giant cluster. So the output isn't a single clustering of the data, but rather a hierarchy of potential clusterings. How you choose which clusters to merge and how we find the "happy medium" clustering in the hierarchy determine the specifics of the algorithm.

An advantage of agglomerative clustering over k-means is that (depending on how you choose to merge your clusters) your clusters can be any size or shape. A major disadvantage though is that there's no natural way to assign a new point to an existing cluster. So it is more useful for gaining insights from a large but static corpus of data, rather than setting up software to handle new data in real time.

5.4.3.4 Advanced Material: Evaluating Cluster Quality

Algorithmic methods to evaluate the outcome of clustering come in two major varieties. First there are the supervised ones, where we have some ground-truth knowledge about what the "right" clusters are, and we see how closely the clusters we found match up to them. Then there are the unsupervised ones, where we think of the points as vectors in d-dimensional space and measure how distinct clusters are from each other.

Silhouette Score

The silhouette scores are the most common unsupervised method you'll see, and they are ideal for scoring the output of k-means clustering. It is based on the intuition that clusters should be dense and widely separated from each other, so like k-means it works best with dense, compact clusters that are all of comparable size. Silhouette scores aren't applicable to things like a doughnut-shaped cluster with a compact one in the middle.

Specifically, every point is given a "silhouette coefficient" defined in terms of

- a = the average distance between the point and all other points in the same cluster
- b = the average distance between the point and all other points in the next-closest cluster

The silhouette coefficient is then defined as

$$s = \frac{b - a}{\max(a, b)}$$

The coefficient is always between -1 and 1. If it is near 1, this means that b is much larger than a, i.e. that a point is on average much closer to points in its own cluster. A score near 0 suggests that the point is equidistant from the two clusters,

Figure 5.17 The indicated point is closer to the middle of the other cluster than its own, because its cluster is large and diffuse. This will give it a poor silhouette score, even though it's actually properly labeled.

i.e. that they overlap in space. A negative score suggests that the point is wrongly clustered.

The silhouette score for a whole cluster is the average coefficient over all points in the cluster.

The silhouette score is far from perfect. Look at Figure 5.17, for example. One cluster is much larger than the other and they are very close. The point indicated is clearly in the correct cluster, but because its cluster is so large it is far away from most of its cluster-mates. This will give it a poor silhouette score because it is closer, on average, to points in the nearby cluster than its own cluster.

However, the silhouette score is straightforward to compute and easy to understand, and you should consider using it if your cluster scheme rests on similar assumptions.

Rand Index and Adjusted Rand Index

The Rand index is useful when we have knowledge of the correct clusters for at least some of the points. It doesn't try to match up the clusters we found to the right ones we know, because we usually don't have a mapping between the clusters we found and the correct clusters. Instead it is based on the idea of whether two points that should be in the same cluster are, indeed, in the same cluster.

A pair of points (x, y) is said to be "correctly clustered" if we put x and y in the same cluster and our ground truth also has them in the same cluster. The pair is also correctly clustered if we put x and y in different clusters and our ground truth also has them in different clusters. If there are n points for which we know the right cluster, there are $n(n - 1)/2$ different pairs of such points. The Rand index is the fraction of all such points that are correctly clustered. It ranges from 0 to 1, with 1 meaning that every single pair of points is correctly clustered.

The problem with the Rand index is that even if we cluster points randomly we will get some pairs correct by chance, and have a score greater than 0. In fact, the average score will depend on the relative sizes of the correct clusters; if there are many clusters and they're all small, then most pairs of points will, by dumb luck, be correctly assigned to different clusters.

The "adjusted Rand index" solves this problem by looking at the sizes of the identified clusters and the size of the ground truth clusters. It then looks at the range of Rand indices possible given those sizes, and scales the Rand index so that it is 0 on average if the cluster assignments are random, and still maxes out at 1 if the match is perfect.

Mutual Information

Another supervised cluster quality metric is mutual information. Mutual information is a concept from information theory; it is similar to correlation, except that it applies to categorical variables instead of numerical ones. In the context of clustering, the idea is that if you pick a random datapoint from your training data you get two random variables: the ground-truth cluster that the point should be in, and the identified cluster that it was assigned to. The question is how good a guess you can make about one of these variables if you only know the other one. If these probability distributions are independent, then the mutual information will 0, and if either can be perfectly inferred from the other, then you get the entropy of the distribution.

5.5 Learning as You Go: Reinforcement Learning

The machine learning techniques we have discussed so far get applied to an existing dataset. In many business situations analyses are done after-the-fact, and any models that get trained on the data are static until a new analysis is performed.

Reinforcement learning is a different machine learning paradigm, where decisions are made in real time and you get to see the results. The initial decision-making strategy might be quite crude, but the algorithm learns by trial-and-error as more data becomes available and (hopefully) converges on a strategy that works effectively.

Reinforcement learning is an important technique when you are deploying a piece of software that aims to be "intelligent" but for which there is no training data at first. It is laborious and error-prone to have human data scientists hurriedly re-analyzing the data as it comes in so that they can constantly deploy updated models. Reinforcement learning makes it possible to have the system simply run itself, learning and improving without a human in the loop. Of course you still want data scientists to carefully inspect the system's performance after the fact and look for patterns that the algorithm might have missed – there is no substitute for human insight.

The key concept in reinforcement learning is the "exploration-exploitation tradeoff": we want to balance trying out different strategies to see what works against milking the strategies that have worked the best so far. A typical reinforcement learning algorithm will start by always making random decisions. As the data starts to accumulate it will sometimes make decisions at random, and other times it will make its best-guess at the optimal decision.

Reinforcement learning is necessarily a cooperative process between your algorithm and a real-world situation. You cannot use historical data to try out different models, because the data would have looked different if you had made different decisions. This means that reinforcement learning is a relatively high-stakes situation: when you pick a model to use you are not only putting it into production, you are also barring yourself from seeing how other models would have fared.

5.5.1 Multi-Armed Bandits and ε-Greedy Algorithms

The simplest reinforcement learning model is called the "multi-armed bandit." Imagine that you are in front of a slot machine with multiple levers, where each lever will give a numerical reward (or punishment) if you pull it. A lever's reward is generated randomly; it might draw your reward from a normal distribution with some mean and standard deviation, flip a coin to decide what your reward will be, or some other method. Here is the hitch: you don't know the parameters of this randomness (like the bias on the coin or the mean of the normal distribution) and they vary from lever to lever. If you knew which lever had the highest average reward you would pull that one exclusively, but you are forced to make a lot of random pulls to figure out which levers are the best. Since your rewards are random you would need multiple pulls for a given lever in order to be confident of how valuable it is.

There are many situations where a multi-armed bandit would be a reasonable way to model the world. The "levers" could be different ads that you show visitors to a website with the hope that they will click on them. The levers could even be different machine learning models that are used to tailor the ads to each user. In some cases the choice of levers becomes an interesting business decision in its own right: if you are looking at different price points for a good that you are selling you will have to choose a handful of prices to experiment with.

The key assumption in the multi-armed bandit is that all of the pulls are independent. Which levers you have pulled previously, or how many pulls you have made total, has no bearing on the potential outcome of the next pull. This assumption makes sense in a situation like showing ads on a website, because generally there is no coordination between the visitors. It can break down though if you are working with a single system that changes over time.

The simplest reinforcement learning algorithm for a multi-armed bandit is called ε-greedy. The Greek letter ε – pronounced "epsilon" – is commonly used in mathematics to denote a number that is small. In this case ε measures how much we value exploration rather than exploitation. At every point in time you pull a random lever with probability ε, and with probability $1 - \varepsilon$ you pull whichever lever has given the best average returns so far. A higher value for ε means that you find the best lever faster, but at the cost that you are regularly pulling sub-optimal levers. In many production systems ε slowly decays to 0 over time, so that you are eventually pulling only the best lever.

5.5.2 Markov Decision Processes and Q-Learning

A key limitation of the multi-armed bandit model is that all of the pulls are completely independent of each other. This works well if every "pull" is a completely different real-world event, but it breaks down if we are interacting with a single system that changes over time.

A more sophisticated alternative is called a "Markov decision process" (MDP). An MDP is like a multi-armed bandit, except that the slot machine is in one of several different internal states. Pulling a lever generates a reward, but it also changes the state of the machine. The behavior of each lever depends on what state you are in. This introduces the concept of delayed gratification: you may want to pull a low-reward lever that puts the machine into a state that will be more profitable in later pulls. In the multi-armed bandit the value of a lever was the average of the reward it gave you. In an MDP it will be the average reward *plus* a measure of how valuable all your future rewards will be.

There is an algorithm called Q-learning that is used to measure how valuable every lever is, in every state of the system. The value of a lever is defined to be the average of the reward you get by pulling it, plus the "discounted" sum of all rewards you will receive in future pulls. When I say "discounted" I mean that the value of a reward decays exponentially with the number of steps until you get it; for example, a reward might be worth 90% of its value if you get it one step later, $90\% \times 90\% = 81\%$ if you get it in two steps, and so on. Adding up the current reward, plus the discounted sum of all future rewards, puts all parts of a lever's value into a single number. The same idea is called "discounted cash flow" in economics. The discount factor (0.9 in our example) is a parameter that gets set manually and reflects how much you value future rewards. If you set it to zero then this all reduces to a multi-armed bandit.

The Q-learning algorithm has a table that estimates how valuable every lever is for every state. Every time you are in a state and you pull a lever, the estimated value of that lever and state gets updated. The update reflects both the reward you receive and the previously-estimated value of the state you now find yourself in. Under certain reasonable constraints Q-learning is guaranteed to work correctly: your estimated values will converge to the actual values of each pull as defined using the discounted rewards. Crucially Q-learning does *not* tell you which lever you should actually pull; it estimates each lever's value, but the actual decision is often made in an ε-greedy fashion.

Glossary

Area under the ROC curve	Literally the area under the ROC curve, as a method for assessing the performance of a classifier across all its different classification thresholds. It is the probability that if you take two random points with correct labels 0 and 1 the classifier will give a higher score to the point labeled 1.
Curse of dimensionality	A variety of mathematical pathologies that occur in high-dimensional data, often having to do with the fact that points on average get farther apart than in lower dimensions.
Dimensionality reduction	Techniques that turn d-dimensional numerical vectors into k-dimensional vectors while still trying to conserve as much of the data's meaning as possible. It is often a practical necessity for dealing with high-dimensional data in machine learning contexts.
Exploration-exploitation tradeoff	This is the central tension in reinforcement. On the one hand you want to try out new strategies to see how well they work. On the other hand you want to milk the strategies that are the most promising.
Factor analysis	A statistics technique that analyzes correlations between different features in data to determine a smaller number of underlying factors that can explain most of the variation in the data.
Gaussian mixture models	A clustering technique that allows for overlapping clusters.
k-means clustering	The most famous clustering algorithm, which is prized for its conceptual simplicity and easy implementation.
Multi-armed bandit	A simple reinforcement learning model where there are several levers that give a reward when pulled. The rewards are

	random, and you don't know the parameters of the randomness.
Markov decision process	A more sophisticated version of the multi-armed bandit, where the slot machine has some internal state. Pulling a lever doesn't just generate a reward – it also changes the state of the machine. The behavior of a given lever depends on what state the system is currently in.
Principal component analysis	The most popular dimensionality reduction technique. It projects points from a high-dimensional space linearly onto a smaller subspace that contains the maximal amount of the data's variation.
Q-learning	An algorithm in reinforcement learning that estimates the value of every lever you can pull for every state in a Markov decision process.
Reinforcement learning	A field of machine learning where decisions are made in real-time, and the resulting data is used to make better decisions in the future.
ROC curve	A graph that plots the performance of a classifier over all possible values of its classification threshold.
Scree plot	A plot of how much variability in a dataset is explained by each of its principal components, which is used for gauging what the "real" dimensionality of the data is.
Supervised learning	Machine learning where we have data that has labels (which may be numbers) associated with the datapoints, and we train an algorithm to be able to predict those labels on future data.
Unsupervised learning	Machine learning where there are no labels associated with the training data, but instead we are trying to understand its latent structure. Clustering and dimensionality reduction are examples of this.

6

Knowing the Tools

This chapter will give an overview of specific software tools that data scientists use, and discuss when they are applicable and what the tradeoffs are between them. I will not teach you how to use them yourself, except to give an optional crash course in database queries.

Choosing software tools is an aspect of analytics in which business people play a critical role, because it extends well outside the sphere of an individual data scientist getting their word done. Some tools are not free and will require large checks to be written. Tools get shared across teams and organizations, models developed with one tool may need to be deployed on another, etc. And it can be very hard to change your mind later.

Technology evolves rapidly, and business decision makers often face immense pressure to use the latest and greatest tools. I saw a lot of this early on in my career as companies wasted many thousands of dollars on Big Data technologies that were fundamentally ill-suited to the problems they were solving. The goal of this chapter is to give you the background to make those decisions correctly the first time. To that end I will:

- Give you a solid conceptual understanding of the types of tools that data scientists use
- Show you the tradeoffs they involve
- Give you a knowledge of what some of the key options are

6.1   A Note on Learning to Code

You do *not* need to be a computer programmer in order to work effectively with data scientists, understand the problems they are solving, and leverage their work. However first-hand experience with software engineering doesn't hurt, and there are times when it is extremely helpful. I would like to discuss how you can dip your toe into coding if you so choose.

Data Science: The Executive Summary - A Technical Book for Non-Technical Professionals,
First Edition. Field Cady.

There are two approaches you can take. One is to learn to write "real code" – developing scripts and software tools in production languages like Python, JavaScript, and C++. There are many online tools that will allow you to do this, like CodeAcademy.org. This is the way to go if you aspire to be a part-time software engineer or data scientist yourself, and it teaches you what their work is like.

The problem though is that you need to know quite a bit of "real code" before those skills become useful in a business environment. The projects you lean on are trivially simple next to any serious software product, and even a "simple one-off script" is likely to have complexities that you won't anticipate going in. You will need to invest a lot of time before the skill become useful, and even then it's not clear that it will be worth it if your primary interest is on the business side.

The approach that I advocate (and it is also optional) is to learn database queries, which I will discuss in this chapter. They are a simplified form of programming that is designed for processing and summarizing data. Many important questions can be answered with queries that are just a few lines long. You could say that database queries are one step beyond spreadsheet programs like Excel; they operate on tables of data and do mostly the same basic operations, but they allow you to express more complicated logic and they involve many of the same concerns as programming languages (reproducibility, efficiency considerations, etc.).

This chapter will include a relatively in-depth discussion of relational database queries.

6.2 Cheat Sheet

The following table gives a quick summary of what I would call the go-to tools for several of the different tasks data scientists regularly do:

Task	Tool of choice	Popular alternatives	Notes
Scripting	Python	R	
General data analysis	Pandas (Python library)	R	
Machine learning	Scikit-learn (Python library)	R	
Graphs	Matplotlib (Python library)		Matplotlib is ripe for being replaced by something better
Database	MySQL	Any other relational database. They're basically interchangeable for a data scientist	

Task	Tool of choice	Popular alternatives	Notes
Big Data processing	Spark, PySpark		Hadoop used to be popular, but it has been eclipsed
Artificial intelligence	Pytorch, Keras (Python libraries)	Tensorflow	Rapidly evolving field

6.3 Parts of the Data Science Ecosystem

Data scientists use a wide range of software tools, overlapping with software engineers, scientists, statisticians, BI analysts, and domain experts. It can be overwhelming, even for those of us working in the field! For simplicity I will break the tools down into several (not necessarily exhaustive or mutually exclusive) categories. I will define them here, and then go into more detail about each one in a subsequent section.

- *Scripting languages*: These are programming languages that – while they can do any computation and can be used to make large pieces of software – lend themselves to writing small-to-medium sized scripts for one-off tasks. Python is the most popular choice for data scientists, but you also see things like Perl and Ruby.
- *Technical computing languages*: This is a programming language that can be used specifically for mathematical modeling and number crunching. Examples include R, Matlab, and Mathematica. Most technical computing languages can be used as scripting languages too, although you don't see it that often. Rather than a dedicated technical computing language, many data scientists use Python along with several technical computing libraries.
- *Visualization tools*: Visualizing data is an absolutely critical part of data science. Any good technical computing library will include functionality for graphing datasets. There are also many separate tools that support interactive dashboards and the like.
- *Big data*: Big Data refers specifically to software that operates on datasets so large that they are stored and processed in parallel by a cluster of computers, rather than a single machine. They are a specialized tool – indispensable sometimes, and otherwise more trouble than they're worth.
- *Databases*: Databases are designed to store medium-to-large datasets and allow you to do simple analyses with very low latency.

6.3.1 Scripting Languages

The key thing in my mind that differentiates data scientists from BI analysts is that the former need to be able to do any computation. Spreadsheet programs, databases, and graphical data analysis tools are wonderful, but as you start asking more sophisticated questions, you quickly run into questions they were not designed for and are incapable of answering. At this point you need the fine-grained control of a programming language.

Almost all programming language are "Turing complete," a technical term that means any computation that can be done in one language can (at least in principle) be done in any other. Databases and spreadsheet programs are generally not Turing complete – there are fundamental limitations on what you can do with them, and data scientists often run into those limitations.

Within the realm of full-fledged programming languages, there are many options. Some (like C) are designed to give extremely fine-grained control over what goes on in the computer, so that you can get the best performance possible; they require very specialized knowledge and are time-consuming to code in, but they can't be beaten for raw performance. Some (like Java and C#) are designed for creating massive programs (think Microsoft Office) with complex dependencies between their various parts, and they include a lot of overhead that becomes valuable for quality control when you are coding on a massive scale. Scripting languages are designed for getting small-to-medium sized projects out the door quickly and easily. They tend to have many built-in functions and libraries that – while not always efficient or elegant – make it easy to hack together something quick-and-dirty. This isn't to say that the code or the languages are low quality, but just that scripting languages shine in situations where simplicity and flexibility are more important than scalability and efficiency.

By far the most prominent scripting for data scientists to use is Python, and I personally swear by it. I like to joke that "python is the second-best language for anything you want to do" – for any specialized domain there is probably a better language out there (Ruby for web development, Matlab for numerical computing, etc.), but nothing beats it as a general-purpose tool. Given that data scientists do such a wide range of tasks, it becomes a natural choice.

Python is an excellent scripting language, but what really sets it apart as a tool for data scientists is that it has an extraordinary range of libraries for machine learning and technical computation. They are widely used, well documented, and they elevate Python from a nifty scripting language to a world-class general-purpose tool for data processing.

6.3.2 Technical Computing Languages

A technical computing language is a scripting language that is specifically designed for doing mathematical analysis. They will have built-in features for efficient number-crunching, common mathematical operations, and ways to visualize and interact with the data. Python can function as a technical computing language because of the excellent libraries that it has, but many other options exist that were specially designed for this purpose.

The most popular dedicated technical computing language among data scientists is an open-source option called R. The data science community is largely split between the people who primarily use R and those who prefer Python. Python has a small lead that is steadily growing; what I always tell people is to use the one they already know, but if they don't know either then Python is the one to learn.

An important thing to note is that different industries tend to be quite partial to different tools, even when the work itself is similar. While self-styled "data scientists" tend to use Python, R is more common in the statistics community. If you move into the physical sciences and engineering then Matlab becomes ubiquitous. Especially if you are looking at people who are early in their career, their choice of language tells you less about how good they are at the work of a data scientist and more about where they learned the trade.

I'll now discuss several technical computing languages worth knowing about in more detail.

6.3.2.1 Python's Technical Computing Stack

Though Python is not itself a technical computing language, I would still like to say a few words about the libraries that let it function as one.

Probably the most standard data science library is called Pandas. It operates on tables of data called DataFrames, where each column can hold data of a different type. Conceptually DataFrames are like tables in a relational database, and they provide many of them same functions (in addition to others). Pandas is extremely handy for cleaning and processing data and for basic analyses. However, it does not generally support more advanced numerical operations.

For machine learning the standard library is called scikit-learn. It supports the standard techniques for both supervised and unsupervised learning and is exceptionally well-documented with many full-length example projects online.

As of this writing the standard visualization package is called matplotlib, but that may be changing. While it has all of the basic functions, you would expect from a visualization library it is widely regarded as the weakest link in the Python technical stack. Among other problems the figures look fairly cartoonish by default. It is losing ground to other libraries like Seaborn (which is actually built on matplotlib but arguably has better default behavior) and web-based visualizations like Plotly.

Under the hood the foundational technical Python library is called NumPy. This is a low-level library that lets you store and process large arrays of numbers with performance comparable with a low-level language like C. From a programming perspective NumPy allows users to create numerical array objects and perform various mathematical operations on them. Most of Python's technical libraries either have NumPy arrays as their inputs and outputs, or they provide abstractions that are really just thin wrappers around NumPy arrays.

SciPy provides a wide range of specialized mathematical functions that operate on NumPy arrays. This is used less in pure data science in my experience but becomes more important if you are using Python for the sorts of tasks that Matlab shines in.

Many data scientists develop their code using a tool called Jupyter Notebooks. Jupyter is a browser-based software development tool that divides the code into cells, displaying the output (including any visualizations) of each cell next to it. It's wonderful for analytics prototyping and is similar in feel to Mathematica.

6.3.2.2 R

R was designed by and for statisticians, and it is natively integrated with graphics capabilities and extensive statistical functions. It is based on S, which was developed at Bell Labs in 1976. R was brilliant for its time and a huge step up from the Fortran routines that it was competing with. In fact, many of Python's technical computing libraries are mostly just ripping off the good ideas in R. But almost 40 years later, R is showing its age. Specifically, there are areas where the syntax is very clunky, the support for strings is terrible, and the type system is antiquated. In my mind the main reason to use R is just that there are so many special libraries that have been written for it over the years, and Python has not covered all the little special use cases yet. I no longer use R for my own work, but it is a major force in the data science community and will continue to be for the foreseeable future. In the statistics community R is still the lingua franca.

6.3.2.3 Matlab and Octave

The data science community skews strongly toward free open source software, so good proprietary programs like Matlab often get less credit than they deserve. Developed and sold by the MathWorks Corporation, Matlab is an excellent package for numerical computing. It has a more consistent (and, IMHO, nicer) syntax than R, and more numerical muscle than Python. A lot of people coming from physics or engineering backgrounds are well-versed in Matlab. It is not as well-suited to large software frameworks or string-based data munging, but it is best-in-class for numerical computing.

If you like Matlab's syntax, but don't like paying for software, then you could also consider Octave. It is an open-source version of Matlab. It doesn't capture all of

Matlab's functionality, and certainly doesn't have the same support infrastructure and documentation, but it may be your best option if your team is used to using Matlab but the software needs to be free.

6.3.2.4 Mathematica

If Matlab is famous for its ability to crunch numbers, then Mathematica is best known for crunching symbols – solving large systems of equations to derive reusable formulas, computing numbers exactly (rather than to some number of decimal places), and so on. It was developed by Stephen Wolfram, a physicist-turned-computer-scientist who is probably one of the smartest people alive. Mathematica is not as common as some of the other entries on this list, and it is not free, but those who like it tend to love it.

6.3.2.5 SAS

Statistical Analysis Software (SAS) is a proprietary statistics framework that dates back to the 1960s. Like R there is a tremendous amount of entrenched legacy code written in SAS, and a wide range of functionality that has been put into it. However, the language itself is very alien to somebody more used to modern languages. SAS can be great for the business statistics applications that it is so popular in, but I don't recommend it for general-purpose data science.

6.3.2.6 Julia

A recent addition to the field of technical computing is Julia, and it remains to be seen how much traction it will ultimately get. Julia has a clean and minimalistic syntax, drawing inspiration from Python and other scripting languages. Its main claim to fame is that while you can develop scripts in it like Python you can also compile those scripts into lightning-fast programs comparable with if you'd written the code in C.

6.3.3 Visualization

Every technical computing language will have graphics capabilities built-in, and these are often enough. If all you need is static image files like JPEGs (usually for use in slide decks and/or written reports), then I usually recommend using those capabilities rather than investing the effort (and sometime money) to learn a new tool.

However, if you want interactive graphics, dashboards that update on a regular basis, or even just things that look gorgeous, then you may want to look at other options. There are too many for me to list here, but I will review some of the ones that are particularly notable or that you're likely to have heard of.

There are several high-level considerations to keep in mind when comparing visualization tools:

- Interactive dashboards are usually powered by a relational database on the backend, but many of the people who want dashboards don't know how to write database queries. Some tools are aimed at these non-technical users and provide visual interfaces for selecting what data should be graphed. This is a great way to democratize access to the data, but data scientists often need the fine-grained control and/or the complexity that you can only get by writing your own queries. Very few tools do an effective job at serving both audiences.
- Data visualization tools used to be stand-alone pieces of software, like Excel, but an increasing number of them are designed to run in a web browser. This makes it easier to do things like embed the graphics in a separate webpage or pull in content from the Internet, and it allows for products that run on websites rather than having to be installed on every computer that uses them.

6.3.3.1 Tableau

Tableau is the industry standard for gorgeous looking dashboards and graphics that you can display in any boardroom. It provides a visual interface that allows non-technical users to select the data they want and use it to create stunning graphics. But if you start asking complicated questions or requiring more fine-grained control, you quickly run into its limitations. For this reason Tableau is generally not a suitable tool for data scientists to use, unless it is purely for purposes of presenting the final results.

6.3.3.2 Excel

In my mind Excel deserves more credit than it gets. Very few programs do such a good job of being accessible to non-technical users for simple analyses while also providing relatively sophisticated functionality to power users. Spreadsheet programs are not a core tool for data scientists, but they are handy for simple analyses and invaluable for making results accessible across a team.

6.3.3.3 D3.js

JavaScript is the programming language that controls the behavior of a website in a browser, and D3.js (standing for "Data-Driven Documents") is a JavaScript library that allows the browser to feature beautiful, efficient graphs of data. D3.js is free and open-source, but it gives very fine-grained control and can be finicky to use. As such it is very popular among analytics firms who use it as a key building block in their web-based analytics applications. Many of the paid visualization tools you might use are built on D3.js under the hood.

6.3.4 Databases

Databases are often a central piece of a data science system for the simple reason that that's where the engineers are storing the data that gets analyzed. Oftentimes the data's main purpose is how it is getting used in a production application, and data scientists are obliged to use whatever technologies are best for the product.

However, even in situations where the entire system is designed for data analysis, databases often play a critical role. There are two main use cases:

- Oftentimes there is enough data that it isn't all being processed at the same time, or would be impractical to process in the form in which it is stored. A database can serve as the central repository of data – perhaps running on a separate server that is shared by a number of data scientists – and each user pulls down only what they need for their current work. A cloud-based shared storage option also allows a simpler version of this, where the data scientist can pull down only the files that they need. But a database allows for filtering out extraneous columns, selecting all rows of a table that satisfy some sophisticated criteria, and computationally intensive pre-processing steps like joins and aggregations – all as a pre-processing step before the data gets to the data scientist's box.
- Some analytics applications require interactivity, like a graph where you can adjust date ranges or the granularity of a time series. Applications like this require low-latency re-computing of the numbers to be plotted, and this is typically done via query to a database.

If there is a database dedicated to the needs of data scientists, it is often periodically refreshed with data from the production servers. In such cases the analytics box is usually a relational database so as to enable joins and aggregations, which are often not needed in the production system itself.

6.3.5 Big Data

"Big Data," as the term is used today, is a bit of a misnomer. Massive datasets have been around for a long time, and nobody gave them a special name. Even today, the largest datasets around are generally well outside of the "Big Data" sphere. They are generated from scientific experiments, especially particle accelerators, and processed on custom-made architectures of software and hardware.

Instead, Big Data refers to several related trends in datasets (one of which is size) and to the technologies for processing them. The datasets tend to have two properties:

1. They are, as the name suggests, big. There is no special cutoff for when a dataset is "big." Roughly though, it happens when it is no longer practical to store or process it all on a single computer. Instead we use a cluster of computers, anywhere from a handful of them up to many thousands. The focus is on making our processing scalable, so that it can be distributed over a cluster of arbitrary size with various parts of the analysis going on in parallel. The nodes in the cluster can communicate, but it is kept to a minimum.

2. The second thing about Big Datasets is that they are often "unstructured." This is a terribly misleading term. It doesn't mean that there is no structure to the data, but rather that the dataset doesn't fit cleanly into a traditional relational database like SQL. Prototypical examples would be JSON blobs, images, PDFs, HTML documents, Excel files that aren't organized into clean rows and columns, and machine-generated log files. Traditional databases pre-suppose a very rigid structure to the data they contain, and in exchange they offer highly optimized performance. In Big Data though, we need the flexibility to process data that comes in any format, and we need to be able to operate on that data in ways that are less pre-defined. You often pay through the nose for this flexibility when it comes to your software's runtime, since there are fewer optimizations that can be pre-built into the framework.

Big Data tools are widely and erroneously viewed as a panacea that revolutionizes all aspects of data analysis. In reality the tools are specialized to handle the particular computational challenges of massive unstructured datasets, and they trade off ease-of-use and some functionality in exchange for it. So you should only use Big Data tools if your data requires it, and even then expect to use them in conjunction with more traditional technologies. It is common to pre-process a dataset using Big Data technologies to get it down to a more manageable size, after which traditional tools are used for finishing the analysis.

6.3.5.1 Types of Big Data Technologies

Big Data tools can roughly be broken down into data processing and data storage. Both of these distribute data and computational work across multiple computers, but they differ in the use cases they serve. Newer generation Big Data tools often technically serve both use cases, but have a strong preference for one over the other.

Data processing tools are of most interest to data scientists. They are designed to do large-scale analyses that cover a large dataset, typically condensing it into summary statistics or transforming it into a new dataset that gets consumed by another process. The processing is done in parallel by multiple computers as much as possible, but usually the computers must also coordinate and exchange data with each other. They generally operate using the "map-reduce" paradigm, which will be discussed in Section 6.3.6.

Data processing tools are usually used either for analytics that isn't in a product, or for batch jobs that may take several hours to run.

Data storage tools are more common as the cloud-based backend of a software product. Imagine that you have thousands of users for a phone app, each of whom has a profile that might contain a lot of complex data. You need to be able to access and edit those data in real-time as multiple users use the service. On the other hand there is usually no need to combine the records for multiple users into a single job.

6.3.5.2 Spark

Spark is the leading Big Data processing technology these days, having largely replaced traditional Hadoop map-reduce. It is usually more efficient, especially if you are chaining several operations together, and it's tremendously easier to use. From a user's perspective, Spark is just a library that you import when you are using either Python or Scala.

The central data abstraction in PySpark is a "resilient distributed dataset" (RDD), which is just a collection of python objects. These objects are distributed across different nodes in the cluster, and generally you don't need to worry about which ones are on which nodes.

A Spark job operates on RDDs, performing various operations that turn them into other RDDs and storing them out appropriately. However, RDDs that are small enough to fit onto a single node can also be pulled down into local space and operated on with all of the tools available in Python and Scala.

6.3.6 Advanced Material: The Map-Reduce Paradigm

The start of "Big Data" is often dated to 2004, when Google published a paper describing the ideas behind MapReduce (MR), an internal tool they had developed for programming jobs that run on a cluster of computers in a way that doesn't require you to think about what data is on which node in the cluster. An engineer named Doug Cutting quickly implemented an open-source version of MapReduce called Hadoop, which is really what popularized the idea of Big Data. There have been a number of re-implementations of the MapReduce paradigm, both proprietary and open-source.

MapReduce is the most popular programming paradigm for Big Data technologies. It makes programmers write their code in a way that can be easily parallelized across a cluster of arbitrary size, or an arbitrarily large dataset. Some variant of MR underlies many of the major Big Data tools, including Spark, and probably will for the foreseeable future, so it's very important to understand it.

The MapReduce paradigm itself is easiest to understand via an example. Say we have a collection of text documents distributed across the nodes of our cluster, and our goal is to compute how often different words occur across the entire corpus. This can be done in three steps:

1. *The local reduce stage*: Every computer counts the word occurrences for only the documents stored on it. The computer will produce a record for every word it found, saying how often that words occurred. This record is often thought of as a key/value pair, containing the word and its number of occurrences respectively.
2. *The shuffle stage*: Every word that has been found anywhere gets assigned to some node in the cluster. Every record produced in stage 1 gets sent over the network to the node associated with it.

3. *The reduce stage*: Every computer that has received records adds up the counts for every word that it is handling.

In this example the Reduce step - where you take multiple key/values pairs for the same word and add them up into an aggregate - is broken into a local stage (also called a "combine" stage) and an overall stage. This is just to reduce network traffic - you could also have just used the overall stage. The Map stage breaks a document into its component ones and makes a key/value pair for every one of them, even if there are duplicates. The local reduce stage then condenses this into a single pair for every word. Only then, when each computer has distilled its text down to the smallest summary possible, does the shuffle stage and full reduce happen. The process is shown in Figure 6.1.

An MR job takes a dataset – which you can think of just as a distributed collection of data objects like JSON blobs or pieces of text – as input. There are then two stages to the job:

- *Mapping*: Every element of the dataset is mapped, by some function, to a collection of key-value pairs.

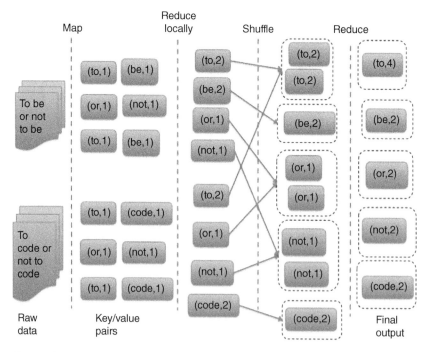

Figure 6.1 The map-reduce paradigm is one of the building blocks of the Big Data movement. The programmer needs only specify what the map and reduce functions are. The job can then be run on a cluster of any size and without regard to which pieces of data are stored on which cluster.

- *Reducing*: For every unique key a "reduce" process is kicked off. It is fed all of its associated values one-at-a-time, in no particular order, and eventually it produces some outputs.
- Sometimes a local reduce is included, but usually it is the same operation as main reducer.

And that's all there is to it: the programmer writes the logic for the mapper function, and they write the logic for the reducer, and that's it. No worrying about the size of the cluster, what data is where, etc.

6.4 Advanced Material: Database Query Crash Course

In an RDB, a dataset is represented by a table with rows (often unordered) and columns. Each column has a specific datatype associated with it, like an integer, a timestamp, or a piece of text (typically with a known maximum length for performance reasons). An RDB has a "query language" associated with it, which lets users specify which data should be selected and any pre-processing/aggregation that should be done before it is returned. The database is structured so that those queries can be answered extremely efficiently.

The SQL family is a large class of relational databases that have nearly identical query languages (even if the internals of the systems are often quite different, optimized for supporting different use cases). Of these, MySQL is the most popular open-source version. This section will be based on MySQL, but know that almost everything translates over to the rest of the SQL family. And in fact it will apply well outside of that even; SQL syntax is ubiquitous, and many data processing languages borrow heavily from it.

6.4.1 Basic Queries

The data in a MySQL server consists of a collection of tables, whose columns are of known types. The tables are organized into "databases" (yes, the term is being abused here). A database is just a namespace for tables that keeps them more organized; you can switch between namespaces easily or combine tables from several namespaces in a single analysis.

A simple MySQL query will illustrate some of the core syntax:

```
USE my_database;
SELECT name, age
FROM people
WHERE state='WA';
```

The first line is saying that we will be referring to tables in a database called my_database. Next, it is assumed that you have a table called "people" within my_database, with columns name, age, and state (and possibly other columns). This query will give you the name and age of all people living in Washington state. If we had instead said "SELECT *", it would have been shorthand for selecting all of the columns in the table. This selection of rows and columns is the most basic functionality of MySQL.

It is also possible to omit the USE statement and put the name of the database explicitly in your query:

```
SELECT name, age
FROM my_database.people
WHERE state='WA';
```

Aside from just selecting columns, it is also possible to apply operations to the columns before they are returned. MySQL has a wide range of built-in functions for operating on the data fields that can be used in both the SELECT clause and the WHERE clause. For example, the following query will get people's first names and whether or not they are a senior citizen:

```
SELECT SUBSTRING(name,0,LOCATE(name,' ')), (age >= 65)
FROM people
WHERE state='WA';
```

Note the somewhat clunky syntax for getting somebody's first name. The term LOCATE(name,' ') will give us the index of the space in the name, i.e. where the first name ends. Then SUBSTRING(name,0,LOCATE(name,' ')) gives us the name up to that point, i.e. the first name. In Python it would have made more sense to split the string on whitespace, and then take the first part. But doing so calls into existence a list, which can be an arbitrarily complex data structure. This is anathema in performance-focused MySQL! MySQL's functions generally don't permit complex datatypes like lists; they limit themselves to functions that can be made blindingly efficient when performed at massive scale. This forces us to extract the first name in this round-about way. Table 6.1 summarizes a few of the more useful functions.

6.4.2 Groups and Aggregations

Besides just selecting rows/columns and operating on them, it is also possible to aggregate many rows into a single returned value using a GROUP-BY statement. For example, this query will find the number of people named Helen in each state.

```
SELECT state, COUNT(name)
FROM people
GROUP BY state
WHERE first_name='Helen';
```

Table 6.1 These functions – which are present in most SQL-like languages – take in a single value and return a value.

Function name	Description
ABS	Absolute value of a number
CONCAT	concatenate several strings
CONVERT_TZ	Convert from one time zone to another
DATE	Extract the date from a datetime expression
DAYOFMONTH	The data in a month
DAYOFWEEK	Day of the week
FLOOR	Round a number down
HOUR	Get the hour out of a datetime
LENGTH	Length of a string in bytes
LOWER	Return a string in lowercase
LOCATE	Return the index of the first occurrence of a substring in a larger string
NOW	The current datetime
POW	Raise a number to a power
REGEXP	Whether a string matches a regular expression
REPLACE	Replace all occurrences of a particular substring in a string with a different substring
SQRT	Square root of a number
TRIM	Strip spaces from both sides of a string
UPPER	Return the upper-case version of a string

The COUNT() function used here is just one of many aggregator functions, which condense one column from many rows into a single value. Several others are listed in Table 6.2:

It is also possible to group by several fields, as in this query:

```
SELECT state, city, COUNT(name)
FROM people
GROUP BY state, city
WHERE first_name='Helen';
```

A final word about basic queries is that you can give names to the columns you select, as in this query:

```
SELECT state AS the_state,
    city AS where_they_live,
```

Table 6.2 Common SQL aggregation functions.

Function name	Description
MAX	Max value
MIN	Min value
AVG	Average value
STDDEV	Standard deviation
VARIANCE	Variance
SUM	Sum

```
        COUNT(name) AS num_people
FROM people
GROUP BY state, city
WHERE first_name='Helen';
```

You could also have only renamed some of the columns. This re-naming within the SELECT clause doesn't have any effect if all you're doing is pulling the data out. However, it becomes extremely useful if you are writing the results of your query into another table with its own column names, or if you are working with several tables in the same query. More of those later.

6.4.3 Joins

Besides grouping, the other key ingredient in the query language is the ability to join one table with another. In a join, several tables are combined into one, with rows from the input tables being matched up based on some criteria (usually having specified fields in common, but you can use other criteria too) and then joined into a single wider row. Joining is illustrated in this query, which tells us how many employees of each job title live in each state:

```
SELECT p.state, e.job_title, COUNT(p.name)
FROM people p
JOIN employees e
ON p.name=e.name
GROUP BY p.state, e.job_title;
```

There are two things to notice about the new query. First there is a JOIN clause, giving the table to be joined with people, and an ON clause, giving the criteria for when rows in the tables match. The second thing to notice is that "people p" and "employees e" give shorter aliases to the tables, and all columns are prefixed by the alias. This eliminates ambiguity, in case columns of the same name occur in both tables.

Every row in `people` will get paired up with every row in `employees` that it matches in the final table. So if 5 rows in `people` have the name Helen and 10 rows in `employees` have the name Helen, there will be 50 rows for Helen in the joined table. This potential for blowing up the size of your data is one reason that joins can be very costly operations to perform.

The query earlier performs what is called an "inner join." This means that if a row in `people` does not match any rows in `employees`, then it will not appear in the joined table. Similarly, any row in employees that does not match a row in `people` will be dropped. You could instead have done a "left outer join." In that case, an orphaned row in `people` will still show up in the joined table, but it will have NULL in all the columns that come from the employees table. Similarly, a "right outer join" will make sure that every row in `employees` shows up at least once.

Outer joins are extremely common in situations where there is one "primary" table. Say you are trying to predict whether a person will click on an ad, and you have one table that describes every ad in your database, what company/product it was for, who the ad was shown to, and whether it got clicked on. You might also have a table describing different companies, what industries they are in, etc. Doing a left outer join between the ads and the companies will effectively just be adding additional columns to the ad table, giving additional information about the company that is associated with a given ad. Any companies that you didn't show ads for are superfluous for your study, and so get dropped, and any ad for which you happen to be missing the company data still stays put in your analysis. Its company-related fields will just be NULL.

6.4.4 Nesting Queries

The key operations in MySQL are SELECT, WHERE, GROUP BY, and JOIN. Most MySQL queries you will see in practice will use each of these at most once, but it is also possible to nest queries within each other.

This query takes a table of employees for a company, performs a query that counts how many employees are in each city, and then joins this result back to the original table to find out how many local coworkers each employee has:

```
SELECT ppl.name AS employee_name,
    counts.num_ppl_in_city-1 AS num_coworkers
FROM (
    SELECT
        city,
```

```
        COUNT(p.name) AS num_ppl_in_city
    FROM people p
    GROUP BY p.city
) counts
JOIN people ppl
ON counts.city=ppl.city;
```

The things to notice in this query are as follows:

1. The sub-query is enclosed in parentheses.
2. We give this result of the sub-query the alias "counts," by putting the alias after the parentheses. Many SQL varieties require that an alias be given whenever you have sub-queries.
3. We used the inner SELECT clause to give the name num_ppl_in_city to the generated column, as described above. This made the overall query much more readable.

Glossary

Big Data	A collection of technologies that use a cluster of computers to coordinate on a single large computation where the data is distributed across the cluster.
Document store	A database that stores "documents" of data that are typically either JSON-like or XML.
Join	A relational database operation that combines to tables into one by matching rows between them and combining them into a single wider row.
MongoDB	A popular open source document store.
MySQL	A popular open source relational database.
NoSQL	A movement that largely overlaps with Big Data and focuses on non-relational databases like document stores.
Outer join	A join in an RDB where rows in a table that gets get matched up still appear in the resulting table. They just have NULL entries in the columns from the other input table.
Relational database (RDB)	The most popular type of database, in which data is stored in tables with known schemas.
Scripting language	A full-featured programming language that lends itself to easy-to-write scripts.

Spark	A popular Big Data technology for processing data.
SQL	A relational database, and associated query language, whose syntax has become nearly ubiquitous in databases.

7

Deep Learning and Artificial Intelligence

"Artificial Intelligence" has replaced Big Data as the buzzword of choice in describing data science. It is a loose term that refers to a collection of different technologies, some of which are actually fairly well-established and others of which are at the bleeding edge of technology. As a rule of thumb AI refers to technologies that do the sorts of tasks that you would normally require a human to do. Multiplying large arrays of numbers quickly doesn't fall into that category, but identifying humans in pictures does.

The key technology at the center of modern AI is "artificial neural networks," a family of extremely powerful machine learning models that can do tasks that were previously unimaginable. Neural networks are also much more complicated than traditional machine learning. Most of this chapter will be devoted to neural networks, including their strengths and weaknesses.

It's important to understand though that a lot of what gets appropriately described as "artificial intelligence" is much older, more primitive techniques that nonetheless often outperform deep neural nets. Using deep learning on a simple problem is like squashing an ant with a sledgehammer. Worse in fact, because the ant often doesn't even die! For some problems deep learning is by far the best game in town, and there is no question that it is the lynchpin for some of the most exciting developments in modern technology. Other times though it does worse than simpler techniques.

7.1 Overview of AI

7.1.1 Don't Fear the Skynet: Strong and Weak AI

I would like to briefly discuss the notion of machines becoming "self-aware" and posing an existential danger to humanity, because it is a concern that weighs on the minds of many people. AI poses many thorny ethical challenges for society, but

Data Science: The Executive Summary - A Technical Book for Non-Technical Professionals,
First Edition. Field Cady.
© 2021 John Wiley & Sons, Inc. Published 2021 by John Wiley & Sons, Inc.

thankfully this is not one of them. While advanced AI systems can do an impressive job of mimicking human behaviors in a wide range of tasks, they do so in ways that are fundamentally different from how the human brain operates. Conflating the two is like mistaking a dolphin for a fish.

If you'll allow me to strain an analogy, with an artificially intelligent system the lights are on but nobody's home. The AI community is focused on getting more lights on, getting them brighter, and maybe setting some mannequins up in the windows. There are no credible ideas about how to get somebody into the building, and serious researchers try to avoid the subject.

A distinction is often made between "strong" and "weak" artificial intelligence hypotheses:

- Strong AI, also called General AI, involves a computer having an honest-to-goodness mind, including self-awareness and whatever else goes into a mind (obviously there is not a rigorous list of what constitutes a mind).
- Weak AI means replicating human-like behavior on a specific task or range of tasks. The range of tasks might be impressively large (say, the ability to identify whether an image depicts a human, a bus, or any of thousands of other objects), but it is strictly less than Strong AI.

In Weak AI there is no pretense of having a "real" mind, and it is generally presumed that the AI system will take a task that typically requires a human with a mind and solve it in some fundamentally different way. If Strong AI means having a fish that lives under the water, then Weak AI means having a dolphin that can hold its breath for a long enough time.

Strong AI is the domain of science fiction writers and philosophers: there are no credible ideas of how to go about it, and many experts suspect that true consciousness is a priori impossible for computers. Weak AI, on the other hand, is what this chapter is all about.

As the limitations of AI technologies become more widely understood they will become, like Big Data before them, just another tool in the toolbox. But they are tools that are re-defining the industry in a way that Big Data never did.

7.1.2 System 1 and System 2

In 2011 the psychologist Daniel Kahneman wrote the book Thinking Fast and Slow, in which he provided a perspective on the human mind that bears comparison to artificial intelligence.

Kahneman (while emphasizing that this is an oversimplification) breaks the mind into two components:

- System 1 (aka "thinking fast") is the collection of unconscious processes that allow us to instinctively interpret the world around us, come up with ideas, and have knee-jerk reactions to things. System 1 is running all the time, weaving all the brain's raw sensory input into a cohesive narrative of what's going on in the world and options for how to react to it. It is not logical or self-critical – System 1 uses heuristics and pattern matching to quickly distill a mountain of sensory input into interpretations that work 99% of the time.

- System 2 (aka "thinking slow") is the conscious mind, which slowly and laboriously makes executive decisions. It takes the suggestions of System 1, sifts through them with a critical eye, and makes final judgment calls. System 2 is lazy, deferring to System 1's conclusions whenever possible rather than going to the trouble of second-guessing them. It is not always rational, but it is capable of step-by-step logic.

The main thrust of Kahneman's book is that System 1 plays a much larger role in our minds than we might believe (or be comfortable with), and System 2 plays a smaller one. Many of our cognitive biases and systematic irrationalities come from System 2 deferring to the judgements of System 1, whose heuristics trade off accuracy for speed (this is a feature of System 1 rather than a bug: you definitely want System 1 to make you startle when you see a lion on the savannah, even if that means you will have some predictable, systematic false alarms!).

The earliest AI systems envisioned the human mind largely as an idealized version of System 2, and they focused on creating logical inference engines that could model System 2's decision making mathematically. It was tacitly assumed that the activities of System 1 would be easy for a computer to do, since they are subjectively effortless for us. There is an urban legend that the AI pioneer Marvin Minsky once assigned a grad student to "solve computer vision" as a summer project!

These early efforts mostly failed, and they fell so short of their hype that the very term "artificial intelligence" was largely discredited for a time. It turns out System 2, while capable of logic, is not just a logical inference engine. But more importantly, researchers dramatically underestimated the importance of System 1's heuristics and brute pattern matching.

The pendulum has now swung the other way. Neural networks – the technology that has brought AI back into vogue – are focused exclusively on learning and applying non-logical heuristics and rules of thumb. They work by fitting complicated mathematical formulas to mountains of data, tweaking the weights until – by hook or crook – those formulas start to work on the training data. This crude approach has worked exceptionally well for guessing what objects are present in an image (the weapon of choice being "convolutional neural networks" [CNN]). In some cases it can assess whether a piece of text is relevant to a question that has been posed (see "recurrent neural networks"). At the bleeding edge of research, some neural nets are starting to combine the two approaches, generating a sentence that describes what's going on in a picture and so on. But at the end of the day, they are really just making suggestions and recommendations, filtering the information down to something that can be ingested and synthesized by a human being with a full-fledged System 2.

7.2 Neural Networks

7.2.1 What Neural Nets Can and Can't Do

My favorite illustration of the limitations of neural networks happened in 2016. Researchers Marco Ribeiro, Sameer Singh, and Carlos Guestrin trained neural network to distinguish wolves from huskies. This is an excellent problem because the differences between the animals consist of a wide range of small hints, ranging from the shape of the ears to the intensity of the eyes. It's exactly the kind of thing that humans are good at picking up on, but that we can't readily describe with equations so it is better to train a neural network.

There was a catch though: the researchers trained on images where the dogs were in grass whereas the wolves were on a background that included snow (pretty typical for each animal, but they did hand-select them). The trained network performed quite well, and many humans who were shown the predictions it made on new images tended to trust that it was indeed distinguishing wolves from huskies. Then they pulled back the curtain to reveal the trick. It is notoriously hard to understand how a neural network makes the judgments that it does, but it is often possible to tell which parts of the image contributed most to the final judgment. The researchers identified these regions, and displayed them to get some insight into how the network was working. Lo and behold, the neural network was essentially cropping the animal out of the picture and classifying based on the background.

In this case the researchers knew they were probably training a bad classifier, because they were well aware of the tools' limitations. In fact they were more interested in ways to assess whether a neural network *should* be trusted (peoples' faith in the classifier evaporated after they were shown which parts of the image it clued in on). But the risk of spurious patterns like this is very real, and stories in the industry abound of researchers spending countless hours and dollars only to get a classifier that makes decisions based on something silly.

There are a couple points to take away from this story:

(1) Neural networks, while powerful, have nothing resembling understanding or common sense. The computer did an excellent job of the problem it was assigned – distinguishing between two different groups of images – but it missed the point of the exercise because it had no conceptual understanding.

(2) In lieu of conceptual understanding, neural networks rely on a mountain of complicated heuristics. Those heuristics may or may not be relevant to the underlying problem we want to solve – they are simply patterns that exist in the data.

This is the key to how many of AI's downsides work: the heuristics that it finds can racist or sexist, they can (as in this case) be an idiosyncrasy of the training data that doesn't generalize, or they can be a numerical hash that is much more complicated than the real-world phenomenon that it correlates with.

7.2.2 Enough Boilerplate: What's a Neural Net?

Recall from the chapter on machine learning that a classifier begins with an array of numbers, called "features," which encode the information about something being classified. The output of the model is an array of numbers called "scores," which indicate the likelihood of the different labels we could apply. The classifier is a really just a mathematical formula that computes the scores as a function of the features, and training the classifier means tuning the parameters of that formula so that the inputs and outputs match the training data. At its core a typical neural network is just a classifier. It differs from other classifiers only in the complexity of the model, and correspondingly in the sophistication of the problem that it can solve.

Neural networks are loosely inspired by the human brain, where every nerve cell will take input from several other cells and then either fire or not. The best way to understand a neural network is with a picture, shown in Figure 7.1.

The network is made of "neurons" (the nodes in the chart) arranged into layers. A neuron takes several numbers as inputs and computes a single number as its output. Most commonly it is a logistic regressor: it multiplies each of its input by a weight parameter, adds them up, and then plugs this weighted sum into a sigmoid function to get its output. In a neural network the outputs of a given neuron are

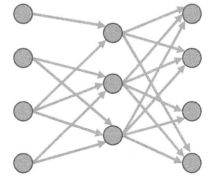

Figure 7.1 A neural network consists of "nodes" arranged into "layers." Each node takes several numbers as input and outputs a single number. The outputs from one layer are used as the inputs for the next. A single node computes it output as a simple function of the inputs – usually a sigmoid.

fed into the next layer as inputs for its neurons, and so on until you get to the final layer. The inputs to the first layer are the raw features, and the outputs of the final layer are our final classifier scores.

Several important points are immediately clear:

1. The "architecture" of the network – how many layers there are, how many neurons are in each layer, and which neurons take which others as input – is as important as the actual parameters that get tuned during training. You may have heard of "Inception V2," "MobileNet," or "AlexNet"; all of these are industry standard architectures that have been found to work well for certain classes of problems. Popular architectures like this often have half a dozen to a couple dozen layers to them, with wide variation in how many neurons there are and how they are wired up.

2. The structure lends itself to identifying complicated features by cobbling together simpler ones. Say that our input is an image, where the input numbers indicate the brightness of all an image's pixels. In one of our layers, there may be a neuron that does a good job of identifying whiskers, another neuron that does a good job of identifying pointed ears, and another that identifies slitted eyes. If all of those feed into a neuron in the next layer, then that one can do a good job of identifying whether there is a cat in the image.

3. When it comes time to train a model, there are lots of parameters to tune, because every neuron in the network will have its own parameters. Popular architectures can have tens of millions of tunable parameters in total, which is usually much more than the number known datapoints to train on, and this makes it extremely prone to overfitting.

7.2.3 Convolutional Neural Nets

Perhaps the most important variant of neural networks is the convolutional neural network, or CNN. CNNs are used in image processing, and they are based on the idea that we often look for a particular feature (which may be composed of several more primitive features) that can occur anywhere in the image.

If you look back at the sketched schematic of a neural net, you can see that every input to the first layer gets treated independently. In the case of an image, these inputs will be the pixels (possibly measuring the light or darkness of the pixel, how much red is in it, or something else depending on the image format), and it is often a bit odd to treat two pixels radically differently. If all of your images were carefully taken in the same way (like mugshots), then it might make sense to treat some pixels differently from others, but for many "real" datasets the pertinent features could occur anywhere in the image (and even in curated datasets, low-level features like lines and geometric shapes could be everywhere).

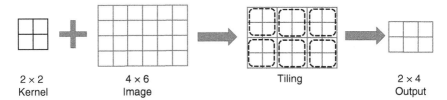

Figure 7.2 Convolutional neural networks are stars in image processing. The first one or several layers of the network are "convolutional": a small grid of numbers called a "Kernel" is tiled over the inputted image and an output is produced for each tile. Usually a Kernel measures the degree to which a particular feature is present in the tile; tiling the Kernel over the image essentially gives us a heat map of which parts of the image match the Kernel well.

In a CNN the first one or several layers are "convolutional." The input to a convolutional layer is an image (technically any two-dimensional array), and the output is a two-dimensional array that measures how strongly each part of the image matches a particular pattern (technically the layer might output several such arrays, each of which measures a different pattern). The pattern is defined by a "kernel," a small image that exemplifies the pattern we are looking for. The kernel will be tiled over the image, and at each location we measure how closely the image matches the pattern at that point. A visualization is shown in Figure 7.2.

The image starts as a large array of numbers, which measure the image's intensity at different pixels. The kernel transforms it into a smaller array of numbers, which measure how much the image matches a particular pattern at each location. Typically several kernels get applied at once. Layering convolutional layers on top of one another allows us to look for patterns-composed-of-patterns, anywhere they might occur in the image.

The size of the kernel and how the tiling is done (sometimes the tiles overlap) are properties of the neural network's architecture – they are chosen by the engineer before training happens. But the kernels are learned by the network during training. Often you can look at the kernel functions to get an idea of what types of patterns the neural network is picking up on.

Essentially all of the best neural networks for image processing are convolutional.

7.2.4 Advanced Material: Training Neural Networks

Strictly speaking a neural network is just a machine learning model, and everything about training machine learning models carries over directly. This includes the importance of training and testing data, the spectre of overfitting, the importance of defining your target variable, and so on. But neural networks have several major differences of degree that make them a different beast to deal with.

7.2.4.1 Manual Versus Automatic Feature Extraction

Feature extraction does not play the central role in deep learning that it does in traditional machine learning. Recall that with most classifiers there is an immense amount of effort that goes into extracting features from the raw data that have some real-world meaning, and those features are what ultimately get plugged into the model. This is usually necessary in order to get good performance out of a model, because the machine learning itself is pretty crude. Feature extraction also has the bonus that we can figure out the real-world patterns that a classifier is picking, which let us sanity check it and guess when it might break down.

In deep learning though, the whole point of a complicated neural network is that the early layers extract features from the raw data, and subsequent layers capture increasingly complex features that culminate in whatever it is you are trying to predict. It's designed for situations where the real-world patterns are complicated enough that it would be impractical to hand-craft good proxies for them.

This is what went wrong in the wolf/dog classification example discussed previously. If researchers had tried to use traditional machine learning to distinguish between dogs and wolves, they would have put together a software that could identify where the animal was in the image and measure physiological characteristics like the length of the snout, the color of the eyes, the shape of the ears, etc. These features – and only these ones – are what would have been fed into the machine learning model for training and testing. They would never have even thought to put in features related to the background of the image, because those are obviously unrelated to the real-world problem they're really trying to solve! In deep learning though, the neural network has the freedom to define whatever features work for the data it is presented with, which are both a blessing and a curse.

The lack of explicit feature extraction does not mean there is no pre-processing to be done – even if features are not explicitly extracted there is often a lot of work to do to make certain that there is not extraneous signal. This could mean, for example, normalizing images to adjust for differences in lighting or replacing jargon in text with standard terms.

7.2.4.2 Dataset Sizes and Data Augmentation

One of the key advantages of careful feature extraction is that you need smaller datasets; if every piece of data can be distilled into just a small handful of highly-meaningful numbers, then very few datapoints are required to discern the relationships that exist between those numbers. Sometimes only a few dozen datapoints are needed to train and test a model if there are only a few, highly-meaningful features extracted.

In deep learning though much more data is needed, because it is necessary to not just train the final classification (i.e. the weights in the last layer of the neural network) but also to deduce the features that will be fed into it (i.e. the weights of all the previous layers). Dataset sizes tend to range in the thousands or tens of thousands. The MNIST dataset – a famous collection of hand-written digits that was used for many of the early successes of deep learning – contained 60 000 training images and 10 000 testing images.

What are you to do if you don't have that much data? A common strategy is called "data augmentation," where a given datapoint is changed in various ways that preserve the correct label – these variations are then treated as new datapoints themselves, multiplying the effective size of the dataset. For example, if you are trying to determine whether an image contains a human face, you can flip it left-to-right and the presence or absence of the face should stay the same. It is also typical (depending on the application) to apply random rotations and re-sizing of images within the context or data augmentation.

The types of data augmentation that are appropriate will depend on what you are studying. Rotating the image by 90° is probably fine if you're looking at pictures taken by a microscope, but will be inappropriate if you're looking for human faces that have a consistent top and bottom to them. Tuning the data augmentation is often a key part of training a robust deep learning model.

I have discussed previously how some people, when faced with a problem of little or no data, mistakenly suspect that it's possible to just simulate data and train a meaningful model based on it. Data augmentation flirts with this idea, but there is a key difference. You need to start with enough data that there are meaningful patterns in it – you cannot use any sort of simulated data to come up with the correct patterns from scratch. But you can use data augmentation to help the neural net weed out many things that are *not* a useful pattern. Flipping images left-to-right ensures that the classifier will not only look for human faces on one side. Randomly rotating images of cells in a microscope ensures that the computer won't find patterns that depend on the orientation and so on.

7.2.4.3 Batches and Epochs

Another issue with training neural networks is that it typically must be done in batches. Usually with traditional machine learning models, the entire training dataset is held in RAM on a single computer, and the best-fit parameters are computed exactly in a single pass.

That does not apply with neural networks for two reasons. Firstly, the training dataset is typically far too big to fit into memory all at once – sometimes by orders of magnitude. The only practical option is to swap subsets of the data in and out of memory. Secondly, the algorithm for training a neural network – called "back-propagation" – is not a clearly defined formula that lets you compute the best-fit weights exactly as a function of the data. Instead it is an iterative algorithm that constantly adjusts the weights, in the hope that they will gradually converge to the best-fit values over the course of many steps.

A typical training session for a neural network is defined by two parameters: the batch size and the number of Epochs over which the training should happen. If your training data consists of 2000 images, you might set your batch size to 100. The dataset would then be randomly partitioned into 20 batches of size 100. The batches would then be loaded into RAM one after another, each one being used to adjust the network's weights. A pass over all of the batches, i.e. the entire training dataset, is a single epoch.

7.2.4.4 Transfer Learning

One of the most important techniques for dealing with small datasets is "transfer learning," which allows you to leverage the fact that somebody *else* had a large dataset that was kind of similar to yours. Say you have a few hundred labeled images of human faces, and you are trying to classify them by gender. This isn't enough data to train a good classifier, but say that you also have an already-trained network that classifies human faces as happy or sad. The early layers in the happy/sad classifier will probably have little to do with sentiment – they will be identifying primitive facial features like eyes and teeth, and those features might be equally useful in guessing gender. In transfer learning you fix all of the early layers in the network, and only use your training data to adjust the weights of the higher-up layers.

Transfer learning has the effect of greatly reducing the number of parameters that you need to learn during training. Large organizations like Google and various research groups pour enormous resources into gathering vast datasets and training very intricate models that perform extremely well on them. When those models get published in the public domain, transfer learning lets us ride their coattails to a vastly better model than our data could ever produce. The key is to find an existing network that was trained on similar images to the ones we are working with, so that the more primitive features are likely to be similar.

Many people – myself included – believe that transfer learning will be the key to the future of deep learning. At the moment it can be a lot of trouble to find a neural network that is suitable for your needs, and success is not assured. But in the future we will begin to converge on a smaller set of extraordinarily powerful networks that possess something functionally akin to "common sense." They will identify the same sorts of patterns that the human mind instinctively hones in on, and bake in many of the same unspoken presumptions that we do. Not because the network actually *understands* any of those things (I'm personally something of a skeptic about the ability of neural nets to ever do that), but because it has so many subtle patterns that it may as well have 90% of the time.

7.2.4.5 Feature Extraction

A particular variant of transfer learning deserves special attention. Recall that the layers of a neural network output increasingly complex numerical features, with the final layer outputting a classification. Typically in transfer learning we would train only the last layer (or the last several) on our data, effectively baking the previous layers' feature extraction into a new classification problem. However, it is also possible to take those features (i.e. the output of some previous layer) and use them directly. They amount to a distillation of a piece of data down to the parts that are most relevant to the problem you are solving.

This is a method for turning a piece of raw data into a numerical vector where the different fields are (hopefully) highly meaningful and distinct from each other. These feature vectors can be applied to other tasks like clustering, estimating the similarity of two points by computing the distance of their vectors, plugging them into a ML model that isn't a neural net, and so on.

7.2.4.6 Word Embeddings

Up to this point we have tacitly assumed that the inputs to a neural network were all numbers. This is fine in the case of images, since the brightness of various pixels is a priori numerical. But for domains like natural language, we must first translate the raw data into a numerical form, and sometimes even go from numbers back to text.

The typical way to do this is with "word embeddings," which maps every word into a vector with a fixed number of dimensions (typically up to a few hundred). We say that the words are "embedded" in the vector space, and we hope this is done in such a way that the words' meanings are captured in the geometry of the embeddings. For example, you will sometimes see researchers boast that their code embeds words like "king" and "queen" such that

$$\overrightarrow{\text{King}} - \overrightarrow{\text{Man}} + \overrightarrow{\text{Woman}} \cong \overrightarrow{\text{Queen}}$$

The mappings are learned from data, and they distill the semantic content of the word into a numerical summary that is suitable for use in machine learning applications.

You may have heard of word2vec, a technique developed by Google in 2013. It is the most famous example of word embeddings, but there are many others. It is a rapidly developing field.

Typically the mapping from words to vectors is customized to every application, by training on a corpus of text that is representative of what we expect to work with. Often this is accomplished by use of a "recurrent neural network" (a type of neural net that is specialized to deal with data that comes in a sequence) that is trained to predict the next word in text based on the words that came before it. The recurrent net is not used for its predictive power; instead it is used for feature extraction as described in Section 7.2.4.5, and those extracted features are treated as a word's embedding.

7.3 Natural Language Processing

Most of the discussion about neural networks has used examples drawn from image processing. The reason for this is that – while neural networks are technically general-purpose machine learning models that can be applied to any numerical data – they have had their greatest success in convolutional networks applied to images.

Natural language processing (NLP) is perhaps the next frontier for deep learning, but so far the results obtained have been less-than-stellar. This section will give you an overview of NLP technologies that blends deep learning with more traditional approaches.

7.3.1 The Great Divide: Language Versus Statistics

There are two very different schools of thought in NLP, which use very different techniques and sometimes are even at odds with one another. I'll call them "statistical NLP" and "linguistic NLP." The linguistic school focuses on understanding language as language, with techniques like identifying which words are verbs or parsing the structure of a sentence. This sounds great in theory, but it is often staggeringly difficult in practice because of the myriad ways that humans abuse their languages, break the rules, and bake assumptions about background information into what they say. The statistical school of NLP solves this problem by using massive corpuses of training data to find statistical patterns in language that can be leveraged for practical applications. They might notice that "dog" and "bark" tend to occur frequently together, or that the phrase "Nigerian prince" is more common in a corpus of emails than in other textual media. Personally I see statistical NLP mostly as a blunt-force workaround for the fact that linguistic NLP is so extraordinarily difficult.

In the modern era of massive datasets (like the web), this divide has become more pronounced, and statistical NLP tends to have the advantage. The best machine translation engines, like the ones Google might use to automatically translate a website, are primarily statistical. They are built by training on thousands of examples of human-done translation, like newspaper articles published in multiple languages or books that were translated. Some linguists protest that this is dodging the scientific problem of figuring out how the human brain really processes language. Of course it *is*, but the bottom line is that the results are generally better and the practitioners are more interested in immediate applications than in basic cognitive science. All of the techniques discussed here will be fundamentally statistical, although they may incorporate some linguistically-oriented pre-processing.

7.3.2 Save Yourself Some Trouble: Consider Regular Expressions

It's worth noting up front that NLP is a relatively sophisticated technique, which means that it can be difficult to use and prone to unusual errors. NLP can be indispensable as a tool, but I have often seen it applied in situations where something simpler would work better. Specifically, you should be prepared to try out a technique called "regular expressions" first.

Regular expressions are a way to precisely describe patterns that occur in text. To take a simple example, the regular expression "A[B-F]*G" is a way to say "any piece of text that starts with A, is followed by any combination of the letters B through F of any length (maybe none of them), and ends with a G". So "ABBEFG," "AG," and "ACCG" are all examples of text that matches the expression; "AXG" and "ABGK" do not match it. Regular expressions are a human-editable way to precisely describe these patterns, and any programming language will have the capacity to parse regular expressions and find occurrences of them in a document.

It's important to emphasize that regular expressions have nothing to do with machine learning (which is the main thing NLP techniques get used for). They are exact rules that get manually crafted rather than probabilistic patterns learned from data. Nevertheless they can often solve a lot of the same problems. For example:

- Say you have a collection of internal messages from within your company's sales team, and you are trying to determine which ones tell you how to contact a new prospective customer. You could use regular expressions to identify phone numbers and email addresses within the messages, and then look only at the messages that contain a phone/email that's not currently in your system.
- Say you have automatically generated transcripts from calls in to your customer service center and you are trying to flag calls where the customer was angry. You could use a list of keywords to find instances of profanity and flag any conversations in which it occurred. But that would be a long list because most profane words have a number of variations (ending with "ing" or "ed," etc.). You could use a regular expression to concisely describe all the variants of a particular profane word and look for that pattern.

In all of these situations the regular expression will work imperfectly. But then again so would a machine learning model, and ML would require the hassle of gathering data and training the model and the reasons why it failed

Table 7.1 Feature of regular expressions.

Feature	Example expression	Matches
[] to mean any one of a group of characters	[Abc42]	A B C
- to indicate a range	[A-F]	A C F
* to indicate arbitrary repetition	A[BC]*	A ACBBBC
{} to give exact counts	[AB]{2}	AB AA BA BB
| to indicate one or the other	A|B	A B
() for nesting expressions	A|(BC*)	A BCCCC

wouldn't always be clear. Regular expressions are easy to create, straightforward to implement, and you know exactly what they are doing.

The biggest problem with regular expressions is that, in practice, the expressions become relatively unwieldy for complicated patterns more quickly than you might expect. It's an interesting case study in human cognition, because it can be easy for us to intuitively "get" a pattern but not be able to really make sense of the regular expression that embodies it. Regular expressions are, apparently, quite different from how the brain represents patterns.

I won't give an exhaustive tutorial on regular expressions, but Table 7.1 shows some of the main features that give them so much flexibility.

7.3.3 Software and Datasets

NLP processing is generally very computationally inefficient. Even something as simple as determining whether a word is a noun could require consulting a lookup table containing a language's entire lexicon. More complex tasks like parsing the meaning of a sentence require figuring out a sentence's structure, which becomes exponentially more difficult if there are ambiguities in the sentence (which there usually are). And this is all ignoring things like typos, slang, and breaking grammatical rules. You can partly work around this by training stupider models on very large datasets, but this will just balloon your data size problems.

There are a number of standardized linguistic datasets available in the public domain. Depending on the dataset, they catalogue everything from the definitions of words, to which words are synonyms of each other, to grammatical rules. Most NLP libraries for any programming language will leverage at least one of these datasets.

One lexical database that deserves special mention is WordNet. WordNet covers the English language, and its central concept is the "synset". A synset is a collection of words with roughly equivalent meanings. Casting every word to its associated synset is a great way to compare whether, for example, two sentences are discussing the same material using different terms. More importantly, an ambiguous word like "run," which has many different possible meanings, is a member of many different synsets; using the correct synset for it is a way to eliminate ambiguity in the sentence. Personally, I think of synsets like the words of a separate language, one in which there is no ambiguity and no extraneous synonyms.

7.3.4   Key Issue: Vectorization

As I discussed in the section on neural nets, a key challenge in dealing with natural language is that machine learning models (including neural networks) operate on numerical vectors, which is a completely different datatype from a piece of text. So the central issue in using machine learning for NLP is how to convert a piece of text into a vector of numbers.

This is a special case of feature extraction, the process of turning our raw data (whatever its format or formats) into meaningful numerical fields. Like other types of feature extraction, this is a complex problem in its own right, but fortunately there are good out-of-the-box solutions because NLP is a well-studied problem. I previously alluded to a very modern approach called word embeddings, where complicated algorithms learn a mapping from word to vector in some high-dimensional space. Usually though the first tool to try is an older, vastly simpler one called bag-of-words.

7.3.5   Bag-of-Words

Probably the most basic concept in NLP (aside from some very high-level applications of it) is that of a "bag-of-words," also called a frequency distribution. It's a way to turn a piece of free text (a tweet, a Word document, or whatever else) into a numerical vector that you can plug into any machine learning algorithm. The idea is quite simple – there is a dimension in the vector for every word in the language, and a document's score in the nth dimension is the number of times the nth word occurs in the document. The piece of text then becomes a vector in a very high-dimensional space.

Most of this section will be about extensions of the bag-of-words model. I will briefly discuss some more advanced topics, but the (perhaps surprising) reality is that data scientists rarely do anything that can't fit into the bag-of-words paradigm. When you go beyond bag-of-words, NLP quickly becomes a staggeringly complicated task that is often best left to specialists.

My first-ever exposure to NLP was as an intern at Google, where they explained to me that this was how part of the search algorithm worked. You condense every website into a bag-of-words and normalize all the vectors. Then when a search query comes in you turn it into a normalized vector too, and then take its dot product with all of the webpage vectors. This is called the "cosine similarity," because the dot product of two normalized vectors is just the cosine of the angle between them. The webpages that had high cosine similarity were those whose content most resembled the query, i.e. they were the best search result candidates.

Right off the cuff, we might want to consider the following extensions to the word vector:

- There's a staggering number of words in English, and an infinite number of potential strings that could appear in text if we count typos and the like. We need some way to cap them off.
- Some words are much more informative than others – we want to weight them by importance.
- Some words don't usually matter at all. Things like "I" and "is" are often called "stop words," and we may want to just throw them out at the beginning.
- The same word can come in many forms. We may want to turn every word into a standardized version of itself, so that "ran," "runs," and "running" all become the same thing. This is called "lemmatization."
- Sometimes several words have the same or similar meanings. In this case we don't want a bag-of-words so much as a bag-of-meanings. A "synset" is a group of words that are synonyms of each other, so we can use synsets rather than just words.
- Sometimes we care more about phrases than individual words. A set of n words in order is called an "n-gram," and we can use n-grams in place of single words.

When it comes to running code that uses bags-of-words, there is an important thing to note. Mathematically you can think of word vectors as normal vectors: an ordered list of numbers, with different indices corresponding to different words in the language. The word "emerald" might correspond to index 25, the word "integrity" to index 1047, and so on. But generally these vectors will be stored as a map from word names to the numbers associated with those words. There is often no need to actually specify which words correspond to which vector indices – it makes no difference mathematically, and doing so would add a human-indecipherable layer in your data processing, which is usually a bad idea. In fact, for many applications it is not even necessary to explicitly enumerate the set of all words being captured. This is not just about human readability: the vectors being stored are often quite sparse, so it is more computationally efficient to only store the non-zero entries.

7.4 Knowledge Bases and Graphs

Another important class of AI technologies is "knowledge systems" or "knowledge graphs." At the moment this is a loose term, so I cannot give you an abstract technical overview in the way that I could for deep neural networks.

Chances are you already use knowledge bases on a regular basis. If you use Google to search for a particular entity (a celebrity, a country, etc.), there is often an infobox on the right side of the search results page, giving some high-level summary information about the target of your search. This is provided by the Google Knowledge Graph, a product used internally at Google that stores (as of 2016) about 70 billion facts about several hundred million distinct entities.

A knowledge base is a way to store information in a structured format that can be understood by a computer. Most knowledge bases are carefully curated and specific to a particular domain of discourse, but in some cases (like the Google Knowledge Graph), they cover a wide range of topics and are generated mostly automatically by crawling a large corpus of documents. A typical knowledge base will contain two components:

- A collection of raw facts about the area of discourse. Typically these are stored in something called an ontology, which I will discuss shortly.
- Some form of inference engine that can process the raw facts to answer questions and draw logical conclusions based on the facts. The implementation of the inference engine will be closely tied to the way the computer stores its facts, and what types of processing that lends itself to.

A key feature of knowledge bases is that – unlike neural networks that might "learn" information by baking it into a complicated mathematical hash – the content can be understood, fact-checked, and edited by humans.

Several of the areas that make notable use of knowledge bases are:

- *Medical science*: Medicine is notorious for having extensive and complicated terminology, including terms and abbreviations that have different meanings in different contexts.
- *Geography*: The layouts of cities, streets, and geographical features are complex and daunting. And yet, once the knowledge base has been formulated, you can query it in fairly simple ways

Both of these examples have a few things in common that underlie the type of situations where a knowledge base might make sense:

- There is a lot of "raw data" that can be used to automatically generate a rough draft of the knowledge base. In these cases it is maps, medical dictionaries, and the like.

- Once you have a rough draft of the knowledge base in-hand, it's the sort of thing where errors can be noted and experts can make manual corrections. The frequency of these errors can be used as a measure of the accuracy of the knowledge base.
- The relationships between different entities can be pretty complicated. This isn't just a nice clean tree-shaped taxonomy. At the same time, there is only a finite number of types of relationships the can exist between entities.
- Critically, the questions you ask from the database generally fall into a few categories that have precise answers.
- The ground truth that the knowledge base is describing is relatively stable. In particular, once you've put in all the effort to correct the rough draft, you can expect it to stay pretty accurate for a while, and you'll mostly have to correct errors as the ground truth itself changes.

Aside from a variety of standardized knowledge bases that exist in the public domain, there is also a trend for companies to create internal knowledge bases. Often these form the core of customer support chatbots, giving users the information they need with a minimum of human intervention. They can also be used as internal tools, condensing a heterogeneous collection of google docs and wikis into a single interface.

Creating and/or maintaining a knowledge base is no trivial matter, and even using a pre-existing one requires coming up to speed on the interface it provides and its strengths and limitations. It's often just not worth the trouble unless you need a particular service that relies on one for its core processing.

Glossary

Bag-of-words	An NLP technique where text if vectorized by turning it into a vector that says how often every word appears in it.
Convolutional neural network	A type of neural network that works especially well for image processing. The early layers are "convolutional" – they tile a fixed kernel across the image and compute how well the image matches the kernel at each point.
Data augmentation	Applying a variety of modifications to a dataset so that there is more total data to train a neural network on.
Knowledge base	A piece of software that combines an ontology of known facts with a logical

	inference engine that can reason about them or provide an easy algorithmic way to explore them.
Lemma	An approximation to the semantic root of a word, capturing its core meaning without the details of how it is being used, such as "run" for the word "running."
Lemmatization	An NLP technique where words are replaced with their lemmas.
Neural network	A family of machine learning models that consist of layers of "neurons," where each layer's input is the output of the previous layer(s).
Ontology	A corpus of facts about some particular domain, stored in a computer-readable format.
Regular expression	A method of expressing precise patterns that can occur in text, which can then be used for tasks like looking for occurrences of that pattern in a document.
Stop words	A collection of words in a language that aren't considered to have useful topical meaning in NLP, and hence are often dropped before applying techniques like bag-of-words.
Transfer learning	Starting with a neural network that has been trained to solve a similar problem to your own, and then re-training only the last few layers on your data.
Word embedding	A mapping from words of text to numerical vectors, which can then be used for various applications like machine learning.
Word2Vec	A popular open-source tool for learning and applying word embeddings.
WordNet	A popular lexical database that groups words into "synsets" all of whose words have a similar meaning.

Postscript

When I first got started as a data scientist, I was learning two cutting-edge programming languages called Pig and Hive. Each of them was a sleek, high-level language that got compiled into a series of map-reduce jobs, which ran on a Hadoop cluster. It was great – you could forget about the awful chore of writing mappers and reducers and focus on the logical structure of your data processing pipeline. Even if Pig and Hive themselves might not stand the test of time, they were the obvious wave of the future; everybody was saying that in 10 years writing your own map-reduce job would be like tinkering with assembly code.

Fast-forward to the present day and Pig and Hive are in the dustbin of history, along with the revolution they promised. Spark has been invented, allowing people write map-reduce jobs in Scala and Python. It turns out that writing map-reduce jobs is easy and fun if you do it in a proper programming language rather than a minimalist monstrosity like Pig. So much for all the predictions...

Data science is not yet at the point where the definitive book can be written – there are too many unknowns. Just how powerful can neural networks be made before we run into fundamental limitations? Will computers become so fast and have so much memory that we can stop trading off flexibility for performance? Is it possible for the educational system to reliably churn out high-quality data scientists, or will they continue to be "unicorns"? The very nature of the discipline – and as a result the future of business in general – hinges on these and other questions.

With that in mind my goal in this book has not been to "teach data science" – that's impossible. Instead I have tried to give you the tools to keep up with the ride! It's not possible to tell which technologies will stand the test of time and which will be forgotten, but I want you to understand the forces behind those changes and to see the writing on the wall when it appears. New analytics paradigms will come to compete with the ones I've outlined in this book, so I want you to understand their strengths and weaknesses. Above all I want you to be able to see past the hype and buzzwords in all situations, so that you can make

Data Science: The Executive Summary - A Technical Book for Non-Technical Professionals,
First Edition. Field Cady.
© 2021 John Wiley & Sons, Inc. Published 2021 by John Wiley & Sons, Inc.

the best possible decisions for your organization with the information available at the time.

Having decried predictions I will now make one of my own: data science will evolve from being a profession to being a skillset. As legacy datasets become more organized, and more user-friendly tools are developed, there is less need for nitty-gritty computer skills to do quality analyses. As pure data scientists like myself finish picking the low-hanging fruit, there will be more need for domain specialists to ask and tackle the truly important questions.

You might use the concepts in this book to help you deal with data science consultants, dumbing your business problems down to the point that an outsider can get something useful done. But the long-term value is with people internal to your organization, or at least your industry. As pure data scientists recede into specialized machine learning models or general-purpose software, the great data scientists of tomorrow will style themselves as "engineers," "biologists," "website designers," or other domain specialists. At that point analytical fluency will pervade every workplace, informing decisions at every level, and the age of data will truly be upon us.

Index

Data Science: The Executive Summary - A Technical Book for Non-Technical Professionals,
First Edition. Field Cady.
© 2021 John Wiley & Sons, Inc. Published 2021 by John Wiley & Sons, Inc.